In Defence of Computer Science

The relationship between computation, the physical machines that perform computation and the subject matter of computations is a subject historically beset by circular definitions and fuzzy boundaries. This book develops a vocabulary that formally splits these three things as ontological concerns. The reader is led through an analysis of difficulties with complete specification in engineering to develop insights into problems with computability definitions. This leads to the proposal of an approach to the specification of useful computations based on zero self-information modulo physical and logical theories, and the development of a set-theoretical characterisation of a 'preservative extensional causal theory' or PECT. It is shown how a solution to Hilbert's sixth problem, a weak form of the second problem and the Church-Turing thesis can be related by means of PECT theories, with a physical theory as the inductive base case. Starting from a consideration of informational and thermodynamic entropy, a theory of quantum gravity is sketched in terms of a PECT theory of 'suprarelativistic nexus dynamics.' This leads to qualitative predictions susceptible to experiment, is not incompatible with some unexplained observations, and has some striking cosmological corollaries if correct. On the notion of a PECT physical theory, the author builds a definition of a logical theory consistent up to a physical theory, and elaborates the nature of prescriptive engineering specifications, with discrete computations a special case. The work concludes in an assembly of the full structure of interplay between these theories through the relationship of transfinite induction and the scientific method, with the suggestion that this points to a mutually inductive topological formulation of physical and logical theories. A final consideration of the second law of thermodynamics gives rise to some interesting implications in the philosophy of mind.

In Defence of Computer Science

from Aristotle to Agile

Samuel R. J. George

EXOCHRON PRESS

EXOCHRON PRESS

PO Box 9646, Dorchester DT1 9JL, England

https://press.exochron.com

In Defence of Computer Science: from Aristotle to Agile
ISBN 978-1-7396928-0-3 (Hardback)

Typeset by the author using X∃LATEX.

Index compiled by Jess Farr-Cox at The Filthy Comma.

Published in England by Exochron Press,
an imprint of Sam George Computer Science Limited.

'Exochron' is a trade mark of Sam George Computer Science Limited.

First published April 2022.

Every effort has been made to ensure that references in this work are correct at the time of publication. However, neither the author nor the publisher can accept any responsibility for the persistence of third party website addresses or the content of referenced material. This work makes scientific and philosophical arguments: it is not intended as an engineering manual and neither the author nor the publisher can accept any liability for loss or damage of any kind arising from such reliance or use.

in memoriam Michael Keith George

Onward, and yet not forward.

—After *Flatland* – Edwin A. Abbott.

Contents

Preface

This work began life as an attempt to tighten up the practice of specification in engineering and computing. However, it became clear that it addressed fundamental issues of cosmology, philosophy and a number of other subjects, so I have decided to present it in the form of a short book. I have tried to strike a balance between developing the material to the extent necessary to present a coherent picture – and the urgency of some of the ideas. I am sure that some mathematical slips and other embarrassing errors will emerge, and I humbly ask the reader's patient indulgence of these. The text ranges over a wide variety of subjects and I am inevitably better informed about some of them than others; I welcome correction on the consequent shortcomings. I have attempted to keep the text lively, even though some of the concepts involved are necessarily dense. The book can be read skipping over the equations and some of the more technical topics: the basic literary thrust of the work should nevertheless be quite accessible. If this Universe works by extraordinary filigreed clockwork, then this book develops the idea that it is – like Leroy Anderson's clock – *syncopated*, and the off-beats perhaps have a special delight as they wash over one's experience of it.

I have attempted to give primary references whenever a clear coherent source for an important concept exists. In the interests of clarity and brevity, I have omitted historiographical discussion for matters that are both very well known and whose emergence is complex, obscure or uninteresting.

Given the sheer scope of material in this book, the reader might – correctly – suspect that I have carefully considered whether such a work is a crackpot enterprise. However, it does appear to be the case that a large number of enigmatic problems can likely only be solved by solving, or at least identifying the key features of a solution of, all of them. I do not believe that this work is half-baked, for a number of reasons. First, the problems it identifies with some current theories and practice seem uncontroversially real. Second, apart from one or two leaps of the imagination, the theories developed here use entirely standard approaches and accepted principles. Third, the physical theory I have proposed submits to the possibility of experiment, observation, disproof or refinement.

While music, poetry and prose have provided much pause for thought in the preparation of this project, in the interests of prompt publication, this edition is presented without chapter epigraphs and the applications for permissions some of these would entail. I hope they may appear in a later edition, which will perhaps illuminate some of the literary intentions more brightly. It is envisaged that a future edition will feature a prologue that will explore the history of some of the concepts touched on in more detail.

It is time to buckle up and begin. It may be quite a ride.

SRJG

Acknowledgements

I am indebted to my Ph.D. supervisor, Greg Michaelson, who patiently tolerated my formative philosophical meanderings in computer science and steered me towards completing my degree with insight, enthusiasm and good humour. In particular, his signpost to the 'Hayes and Jones' debate during my doctoral studies led to a subsequent series of thoughts over the last ten years that culminated in this work. I am grateful too to Art Scott, who read my doctoral thesis and whose interest in its application to adiabatic computing was a further encouragement for me to finish a long-planned paper on the themes of my Ph.D. That paper is still yet to be concluded, but it was in trying to complete it that it seemed necessary to sail into these waters. I did not mean to embark on a nautical adventure.

Chapter 1

Introduction

Two perennial difficulties with computer science are that there is no satisfactory definition of what a computer *is* and the science does not – at least very directly – make testable predictions about Nature.[1] Eliding practical details, an attempt to define a discrete computer usually goes as follows: a computer is a device that furnishes the value of computable functions, which are the functions that could be computed by a universal Turing machine. The definition of a Turing machine itself is ingenious [152], but – for all its mathematical clothing – still essentially appeals to surprisingly elusive notions of 'computability' [152]. The only entirely formal characterisations of such a machine are in terms of other formalisms that themselves cannot substantiate the claim. The Church-Turing thesis[2] – essentially that this equivalence is compelling evidence of its canonicity – while intuitively convincing is ultimately circular and anthropocentric. Computer science is scientific in that humanity is part of Nature and its computations are a valid object of enquiry; it also turns out that a certain definition of a computable function can be a helpful ingredient in axiomatising scientific theories. We

---

[1] We adopt the convention of capitalising 'nature' for the same reason we capitalise 'internet': to refer to the canonically singular instance – not to advance some sort of pantheistic agenda.

[2] The thesis or conjecture has a long development through the modern history of logic but it arguably crystalised in the form we recognise it in [84].

1

shall develop this view later in this book.

A frustration in the idea of 'computability' [152] or 'effective calculabil-
ity' [32] is not at all original. In particular, Gandy and Sieg pushed Turing's ideas deeper towards abstract causation, locality and boundedness
[55, 138], but the resulting axiomatisations can be viewed as tighter restate-
ments of the original idea. They are still somewhat prolix and Ockham's
razor – which one might regard as axiomatising the intuition that Nature is
beautiful[3] – suggests a simpler axiomatisation would be preferred. We can
intuit that such an axiomatisation should be a physical principle, because
attempts to analyse logical processes in terms of themselves lead to well
rehearsed contradictions of self-reference – perhaps canonically in Gödel's
incompleteness theorems [60], but also in Tarski's undefinability theorem
[143] and Turing's answer to the *Entscheidungsproblem* [152]. The intuition
that leads to these theorems (*via* Russell's paradox [129]) lies in long known
conundrums such as the Liar paradox.[4] In this book, we shall use the word
'information' in its normal English or computer science sense as the context
requires, but to avoid ambiguity, shall qualify it by one of our ontological
definitions if we require a more specific meaning that is not obvious in the
context. 'Anthropontic' and 'cosmontic' information refer to encoded data,
divorced from that which they may denote[5] (hypo- and hyperkeimenontic
information respectively). We shall define these terms shortly.

The unsatisfactory nature of the characterisation of 'effective calculabil-
ity' is effectively the same as a similarly frustrating problem in logic: that
the consistency of first-order arithmetic cannot be proved in *itself*, but it
can be proved to the satisfaction of human mathematicians using transfinite induction, as in Gentzen's consistency proof [141], if we accept transfinite induction over a suitable set theory as an axiom. However, *physical* axioms or laws, unlike logical ones, can be tested inductively by experiment: a
proposed escape from circularity based on physical axioms is not dogmatic

---

[3] The idea is so universal such that to attribute it to William of Ockham would be mislead-
ing; it has a history all of its own.

[4] The origins of which are ancient and not known with any certainty.

[5] The modern roots of this idea can probably be traced to [130].

about epistemic or doxastic accounts of semantics but is epistemically humble and corrigible, and can thus be subjected to the scientific method.

In this book we propose to escape from circular deduction by a definition of computability that is based on Shannon self-information.[6] To complete the definition, we shall need to develop concepts of contingent semantic completeness for logical and physical theories. A *minimally adequate* definition of a computable function can be stated informally as follows:

Definition 1.1. A *computable function* is any intensionally[7] describable function whose codomain contains the same as or less than the self-information of the domain and the function's predicate specification.

This is the very opposite of how we usually think of the semantic purpose of computation. We provide an argument that this is equivalent to the idea of 'effectively calculable' by supplying the executable data *decompression* function of the history of the state space of any computational process as the inverse of the function that computes that state space, satisfying Shannon's source coding theorem by construction. Following Lord Kelvin through the lens[8] of an evolutionary conditioning argument, we argue that this is essentially the same statement that physical processes are either reversible or they contain less of the information they started with in the 'forward' direction of the arrow of time. Reversible functions preserve the accessibility of information within the system of interest; irreversible functions erase it; no computable functions create it. If the outcomes of computations have no significance or utility – or if they are already known to an acceptable level of certainty – then computing them is a waste of resources. The epistemic commitment of self-information needs to be Bayesian rather than

---

[6] The term 'self-information' is due to Fano [50, p. 34] (Fano does not hyphenate it), though the very closely related concept of information entropy is due to Shannon [136]; Fano establishes the relationship at [50, p. 42]. The distinction will be relevant to later remarks on temperature in Section 6.1 and the relationship between the cosmontic and hyperkeimenontic explored in Chapter 7.

[7] The intensional *versus* extensional terminology follows in the tradition of Jones [79] and Carnap [27].

[8] Or perhaps looking-glass.

frequentist[9] to avoid later circularity with our ideas about time; this implies
a preference for a conception of probability philosophically aligned with
Cox's system [34] rather than Kolmogorov axioms [86], and our belief in
the plausibility of some aspect of physical reality is a credulity conditioned
by the environment through evolution and experience.

We shall need the following terms, intended as adjectives or nouns (denoting that which is predicated by the adjective). While the definitions are
inevitably a little dense, their full meanings will become easier to grasp as
they are used in contexts established throughout the course of this book. It
is, alas, inevitable that we can only break a circle by starting somewhere:[10]

Definition 1.2. *Cosmontic* – the reality of information *in* the Cosmos as opposed to the Cosmos itself or information about it. It exists extensively in
physical reality but the information is rendered locally inaccessible through
convolution. The structure (such as it is) within that information is not necessarily known, but such as it is, is hyperkeimenontic. The cosmontic can
be thought of as being in the indicative grammatical mood.

Definition 1.3. *Hyperkeimenontic* – the reality in the superset of the physical
that includes the ethereal, knowledge of which is secure up to a willingness
to trust perception of physical reality and acceptance of the induction argument of Chapter 7. The shift in regarding an object as a subject (in the
modern linguistic sense) suggests the belief in a morphism between the hypokeimenontic and the hyperkeimenontic. The ethereally hyperkeimenontic cannot be known with certainty but seems likely to include objects that
might be considered mathematically Platonic. The hyperkeimenontic may
describe Nature but is not *in* it: its grammatical mood is unknown.

Definition 1.4. *Hypokeimenontic* – the reality that exists insofar as one or
more human beings can imagine it in the subjunctive grammatical mood.
The Aristotelian idea that the 'it' of a ὑποκείμενον[11] [8, Categories 1a20], as

---

[9] See [82] for the canonical discussion of the difference.

[10] All of these '-ontic' words refer to a reality, but some of those realities are contingent.

[11] Underlying substance of a thing.

distinct from a predicate, is relaxed in the hypokeimen*ontic* and is thus fused with the idea of κατηγορία[12] in the manner of higher-order logic: our specifications of the anthropontic are themselves hypokeimenontic. The meaning is similar to that of 'subject' before it was complicated by the philosophy of Descartes [39] and the reaction to that philosophy. Definitions of and predicates over the anthropontic are themselves hypokeimenontic; all of our terms can be considered to overlap with the Kantian ideas of 'noumenon' and 'phenomenon' [81, pp. 338–365], which we do not find to be helpful in the present discussion, preferring a less dogmatic account of which particular things actually exist, in a manner more in the spirit of Kierkegaard [83]: it is not taken for granted that any particular hypokeimenontic entity corresponds to anything hyp*er*keimenontic.

Definition 1.5. *Anthropontic* – the reality of something defined as a hypothetical observation of something physically hyperkeimenontic believed to correspond to an approximation of something hypokeimenontic. *Actual* observations in this sense are the extensive obfuscation (or effective erasure) of cosmontic information involved in it encoding an approximation of something *else* that is believed to be physically hyperkeimenontic. Its reality is contingent on the fitness of the model that is the hypokeimenontic object of the belief and its encoding. It includes encodings of hypokeimenontic entities constrained by computable functions of other encoded hypokeimenontic entities, regardless of whether the encoded entities in such codomains *actually are* encoded physically hyperkeimenontically. The anthropontic corresponds with the idea of declaring a variable and can be expressed in the imperative grammatical mood.

The remainder of this book is organised as follows. In Chapter 2 we investigate the nature of engineering specification to arrive at a definition of *complete* specification broad enough to describe and specify both mathematical functions and physical systems. In Chapter 3 we extend the informal discussion of computability to characterise computable functions in terms of

---

[12] Category or predicate.

information content. Chapter 4 addresses causation and temporal specification before a minimal ontology of coordinate systems and complete theories is introduced in Chapter 5. This prepares us for a discussion of specialising the definition of these theories into axiomatisations of scientific theories as well as synthetic mathematical or engineering specifications in Chapter 6, which concludes with a consideration of the axiomatisation of logics. Chapter 6 is longer than the others because we found it was necessary, in order to substantiate our overall argument, to supply in Section 6.1 a sketch of a complete physical theory in the form of a self-contained essay. Finally, in Chapter 7, we identify the nature of a fixpoint structure that defines a relationship between these theories and suggests an equivalence between its inductive step and the fixpoint structure of ε numbers usually relied upon to demonstrate the consistency of logics. This leads us to an interpretation in which a physical theory induces a tight definition of computability through logic, with an implication for the philosophy of mind.

Chapter 2

Nature of engineering specification

One of the most astonishing properties of Computer Science is how *useful* it has proven to be, something which has propelled it from university mathematics departments towards engineering. It is not an entirely happy or complete transition: in a phrase sometimes attributed to Dijkstra, 'Computer Science is no more about computers than astronomy is about telescopes.' [64, p. 4] In order to try and make progress with what computing *is*, we shall therefore start by examining what is often perhaps wrong with it as an engineering discipline, and the conclusion hints at an answer to our main problem. It will provide a heavy hint that the computability definition should be the reverse, essentially thermodynamic one of Definition 1.1.

2.1   In search of correctness

'Correctness' of computing systems can be a slippery idea. The issue is a special case of correctness in engineering more widely, but we shall motivate our discussion with examples from the specification of computing systems. In all cases, it is required to maintain information in a predictable system configuration, notwithstanding the tendency of it to leak away through the vicissitudes of thermodynamics.

Even in the most formal development methods such as Z [77], B and Event-B [2, 3] or vdm [20], correctness can only go as far as using some kind of deductive argument to show that an implementation is consistent with a specification – whether done by refinement, construction or *ex post facto* proof, this is theoretically a closed logical verification exercise, assuming we have a complete enough logic whose soundness we believe in. A more open problem concerns where the specification comes from and whether it suits 'what is wanted' – an issue of validity. The difficulty is that 'what is wanted' encompasses three things: first, it needs some subject matter over which these desires are expressed (whether this subject matter is misidentified is not something that can be addressed by logic). Second, there must be a desire for this subject matter to behave in a particular way: a matter of 'the passions,' to recruit some words of Hume [74, Part III, Section III (pp. 413– 418)]. That much is outside the field of 'reason' and we cannot reasonably seek to reduce it to a mathematical problem. However, where this is mingled with a third issue – inconsistent 'passions' or those that are partially silent on their subject matter, we can invite the owner of those passions (the requirer or specifier) to correct their inconsistency or form a view about the parts of it they neglected to consider. The consequences of not doing so are rather serious, in that any concept of completeness of tests or proof obligations defined in reference to such a truth might be meaningless: it is not necessarily simply an unknown truth, but if it is defined over an understanding that is either incomplete or inconsistent, the truth is instead *non-existent* – the specifier has neither experienced nor articulated the requisite set of mutually consistent passions. To argue the Platonic existence of some such specification is unconvincing since specifications of human passions are by definition artificial such that counterfactual supposition about their content is vacuous. If we reason *ex falso quodlibet*, it would seem such incomplete specifications might compromise a development before it has started.[1]

This idea is hardly original – it is perhaps contained in some over-quoted remarks of Babbage that are nevertheless pertinent and prescient: 'The errors

---

[1] 'Passions' as the prime human mover has a certain resonance with the Pre-Raphaelite sensibility.

which arise from the absence of facts are far more numerous and more durable than those which result from unsound reasoning respecting true data' [10, p. 119], or the even more famous remark about right numbers emerging from wrong inputs [11, p. 67], whose frequent paraphrase as 'garbage in, garbage out'[2] (if we choose to generalise it from the difference engine to the analytical engine[3]) arguably misses part of Babbage's point: it is not about the validity (in some respect) of first-order data, but rather having some coherent idea concerning the effect of the machine into which it is inserted. If the designer (or programmer) of the machine does not understand what they are doing, the user of it may have to rationalise the incompleteness or inconsistency of the resulting device. This hapless individual has in effect their guns spiked before they have started, so the 'second-order' interpretation of this remark identifies an even more hopeless scenario.

To put correctness another way, consider Turing's idea of an 'oracle' [153, p. 172] that knows the output of a function: if this oracle is over-specified, then it is at best arbitrary and at worst capricious; if it is under-specified, then it is Delphic, and the answer is to some extent surprising. Generally, this is a kind of surprise that no engineer wants, especially if development took a long time and consumed considerable resources. *Validation* in engineering is the *ex posteriori* process of checking that the engineering artefacts are *not* surprising, but doing this when otherwise finished risks waste and rework. Such unwanted surprise provides a hint. Shannon information is a measure of 'surprisal' [149, p. 64]: in an engineered system, we do not want any such thing. In our terminology, an engineer may want to use or learn some hypokeimenontic information by executing a computable function,[4] but that information is only being presented in a format that is *useful* in the *opinion* of the specifier, and does not create any information that was not implicit in the starting point.

---

[2] This is widely credited to Raymond Crowley in *Robot Tax Collector Seeks Indications of 'Fudging'* in the Alabama *Times Daily*, 1 April 1963.

[3] *i.e.* consider garbage programs rather than garbage numbers.

[4] The process of giving denotations to computer programs is the business of denotational semantics [134].

This idea has some consequences for artificial intelligence ('AI') [107]: if all computers preserve or are lossy functions of information, then their outputs must just be ways of presenting the data that went into them. If entropy is read in explicitly or through the orderings of parallel processes, then the output is partially a function of random noise. It is therefore impossible to determine with certainty how much of the output encodes genuine truths about the hypokeimenontic from that encoded in its inputs, and how much is an amplification of noise that is either random or indistinguishable to us from random. Machine learning turns this into a second-order phenomenon by generating functions whose codomains lack any complete characterisation beyond a tendency to select for the average.

AI as a kind of heuristic is well suited to *finding* efficient implementations or proofs that can be checked by a deterministic process. This is the idea of the De Bruijn criterion in proof systems [13], which has a corresponding meaning in concrete engineering through the Curry-Howard isomorphism [140]. However, as a general method of discovering information or engineering systems it is conceptually flawed, as it presents a function of what is in effect partially meaningless information[5] in a form in which we give it meaning*ful* semantics: we 'know' the answer but not necessarily all of the question. We are led inexorably to the conclusion that the idea of intelligence artificially materialising as an emergent phenomenon from something we mix with noise and put into algorithms is not artificial intelligence at all but artificial anthropomorphism of the ill-defined anthropontic: this can be a dangerous category error, to use the concept due to Ryle [131] in a way that has a certain irony in the light of what we shall say in Chapter 7.[6]

---

[5] Or to be more precise, if it has a meaning, it is not a meaning that speaks to the question of interest. This relativity of meaning leads us to be wary of pursuing hard distinctions between data and information as a matter of objective truth and justification of belief, which rapidly leads to a descent into the vortex of Gettier problems [57], as described by Floridi [53]. Our broad conclusion in this paper on the extensional topological nature of our physical reality resonates with Floridi's structuralist account of informational reality, but the focus and details of the epistemic commitments are different, and our ideas about correctness differ from those of Floridi, particularly on the issue of validity.

[6] We are not going to be drawn into the 'monist' *versus* 'dualist' debate, as we are not

2.2 Predicate *versus* executable specifications

The merits and demerits of executable and non-executable specifications were argued extensively [54, 63, 68, 154] until the dispute fizzled out. This debate is summarised in [56]. Essentially, after Turner had advanced the idea of using total functional programming languages as a means of specification [154], Hayes and Jones argued that only predicate specifications would avoid elevating implementation details into specifications and permit the expression of non-computable or non-deterministic constructions [68]. Fuchs [54] argued that non-executable specifications can be made so, while Gravell and Henderson [63] concluded that executable specifications could be useful in conjunction with other approaches.

The idea of standalone specification languages is most strongly associated with the lineage of abstractions that includes vdm [20] , Z [77], B [2], Event-B [3] and tla+ [87]. Key to such languages is the refinement methodology that goes with them. Specification starts with the statement of predicates in first-order logic over concreta based on sets. These predicates (specifying invariants) are supposed to be derived from a high-level set of requirements that *under*specifies a system. The specification is gradually decomposed into smaller structures, over which further constraints are introduced by refinement, until a specification in the restricted form of a synthesisable or executable subset is reached. However, at this point, the system is likely to be *over*specified, because it ignores the fact that any number of programs (some trivially different from the one actually arrived at) could do the same job. Of course, the problem is defining mathematically what we nebulously refer to by the phrase 'the same job,' which may not need to be as restrictive as saying that the same input history has to produce the same output history up to Leibniz [128, pp. 12–13] equality.

Refinement methodologies do not provide natural means to show that a specification has found a 'Goldilocks' level of detail: *i.e.*, it is neither over- nor underspecified, but specifies *precisely* those outcomes that accord with an intuitive idea of 'what is wanted.' In standard approaches to specification

---

convinced it can be agreed what those terms mean: see, for example, the analysis of [71].

in reasonably formal development processes, completeness remains an informal notion that is claimed though not formally proved. If we tried to prove this formally by finding an equivalence class of the kind we shall discuss in Section 2.3, and in doing so, found that the putative class included something that seemed with retrospect to belong to the 'what is *not* wanted,' we would have shown the informal assertion of completeness to be mistaken. However, it is not standard engineering practice to look for such formal equivalence classes. Moreover, the restriction of popular refinement methodologies to first-order logics is an obstacle to expressing or proving completeness over specifications in this way, which motivates our use of more general-purpose tools with higher-order logics. It is possible to interpret the impetus for the 'validation' part of the ubiquitous V-model of software development as a mitigation for this deficit of an explicit equivalence class in the original specification.[7]

In recent years Agile development methodologies [14] have increasingly dominated, in which software (although there is no reason why it should be confined to software) is supposed to be developed in rapid iterations in which new 'features' are added. The problem with any verification exercise based on this is that it can lead directly to the fallacy of irrelevant conclusions (*ignoratio elenchi*), even supposing a perfect set of regression tests or proofs. In reality no such set is perfect, as even the regression tests or proof targets have to be updated, and nothing regression tests these. Worse, if we are foolhardy enough to accept an unproven and untrue logical proposition in our requirements, we fall foul of unintentionally reasoning *ex falso quodlibet*, and our efforts are thus undermined by the fallacy of question-begging (*petitio principii*). The fundamental problem is that no Agile methodology requires that a specification be complete – the entire process is one characterised by garbage in, garbage out, even if the development process is executed perfectly. We shall see in the next section that this 'dynamic Agile' chaos is unnecessary if we move the process to a second-order exercise,

---

[7] We observe that when a system such as the B method is used, the specification is protected against false informal claims of *consistency*, because the B method provides a mechanism for dealing with this scenario by producing proof obligations for the consistency of invariants.

in which all effort goes into producing *complete* specifications according to a particular characterisation based on complementarity of functional and equivalence predicates. If specifications are complete, first-order, dynamic Agile and first-order software development is repetitious and pointless and holds those who rely on the fruits of it vulnerable to latent defects, the consequently interminable software update requirements, and hardware obsolescence. Implementation tactics (*i.e.* program-writing programs) with low or no quality control can produce verified programs as long as they type-check in a suitable dependently typed logic, which by the Curry-Howard isomorphism, is exactly the same task as finding constructive proofs of propositions. *Any* program that type-checks against a complete specification is *by definition* correct. There is no reason why we cannot prescribe a tolerated subset of implementation failures either. We now therefore proceed to consider how we might know a complete – neither under- nor over-specified – specification when we see one. This gives rise to a kind of *static* agile process, where the acceptance criteria (or proof targets) are fixed for *any* specification and are therefore not a moving target. If we are concerned about change management in some *finished* specification (as opposed to implementation), the change may be specified as a second-order specification with respect to a first-order migration specification against *third*-order invariants in a similar way to how we specify second-order invariants for first-order specifications. It is time therefore that we stated some requirements for complete first-order specifications, which we shall now do.

2.3 Axiomatising non-modal specification

A necessary and sufficient condition for an executable specification carrying values uniquely from a domain D into a codomain C can be described by the two-place relation F that satisfies

$$\forall d \in D. \ \exists c \in C. \ F(d,c) \wedge (\forall c' \in C. \ F(d,c') \Rightarrow c = c') \tag{2.1}$$

and

$$\forall c \in C. \ \exists d \in d. \ F(d,c) \tag{2.2}$$

The left side of formula (2.1) expresses function totality or completeness, while the right side expresses consistency or soundness. Formula (2.2) ensures that the codomain is a codomain rather than a range – *i.e.*, F is a surjection. F can be promoted to second order if we want a morphism to a function over judgements expressing some interval over a system representation at a lower level of abstraction. F can also be changed to a non-deterministic $\mathfrak{F}$ with an equivalence $c \equiv c'$, but this can be eliminated again if desired by admitting entropy as a part of D. We can also change Leibniz equality to equivalence if required by an embedding in a logic with multiple representations of the same underlying value,[8] although this would be a feature of the embedding, not the function.

The difficulty with our axiom for F is that, for practical computable functions, its structure and interpretation into a big-step mapping expresses the *how* of a specification rather than the *what*: the resulting predicate is over-complicated and unintuitive. If we wish to build specifications by progressive elicitation of constraints, a non-executable specification can be arrived at as the conjunction of a number of disparate requirements of the input/output space:

$$\forall d \in D. \ \forall c \in C. \ P(d,c) \wedge Q(d,c) \wedge R(d,c) \wedge \dots \qquad (2.3)$$

To allow the Hayes and Jones debate discussed in Section 2.2 to converge on some concrete 'Goldilocks' criterion (at which point the dispute evaporates), we must have, in addition to the requirement for a total function:

$$\forall d \in D. \ \forall c \in C. P(d,c) \wedge Q(d,c) \wedge R(d,c) \wedge \dots \Leftrightarrow F(d,c) \qquad (2.4)$$

The predicate of formula (2.3) can never be satisfied if any of the constituent predicates are contradictory. The specification is only meaningfully complete if formula (2.4) holds. Extra predicates that are unnecessary to show formula (2.4) give rise to harmless tautologies, although they might suggest the original requirements were needlessly verbose.

---

[8] Such as in many of the theories of the Coq proof assistant [45, 116].

The complaints that such a requirement is likely to produce are that this property cannot be shown in practical examples because various subsets of D are infeasible, or that the specification only makes demands over equivalence classes such that, for example, any of d_1, d_2, d_3 could give rise to any of c_1, c_2, c_3 and still satisfy the intent of the specification. In either case, we argue that the objections can be overcome: in the first instance, by restricting the domain D, and in the second case, by rephrasing F as a category functor over *explicit* equivalence classes. If the equivalence classes over D and C are Leibniz equality, then the specification is deterministic at the level of abstraction at which it is defined; otherwise it is complete, modulo those equivalence classes. These questions of non-Leibniz equivalence are unavoidable when considering a physical ontology: in a physical weak bisimulation of a discrete specification in which computations are chained together, D must generally be wider than C to accommodate unwanted non-determinism. In a classical control system, finitistic representations in a computer program are only ever approximations of some ultimately unknowable and to some extent un*definable* real number describing some hypokeimenontic entity. Even if using the system to reason about axiomatically discrete objects, the implementation of the system must still be engineered to bisimulate these discrete values modulo the uncertainties in their physical manifestation.

This kind of axiomatisation of a complete specification constitutes an acceptance criterion for a piece of engineering where the executable function is not designed to be an efficient implementation, but rather one that allows automated processes to find any equivalent specification that is efficient according to some desired parameters. We know from mechanised proof systems following the principle of the Edinburgh LCF system [62][9] – such as Isabelle/HOL [115] – that checking an acceptance criterion can be reduced to type checking, a compact procedure that satisfies the De Bruijn criterion.

There is no theoretical reason why the $P, Q, R, ...$ of formula (2.4) cannot be refined by an essentially Agile approach together with constructive proof

---

[9] This is itself based on theoretical work by Dana Scott [133].

that the conjunction of various components induces an executable function, which is what we labelled as 'static agile' in Section 2.2.

The rather glaring problem with our approach to axiomatising specification so far is that we have assumed that we can provide an executable function to demonstrate that a specification is realisable, but we have not yet characterised such a function or provided a method to identify one. We begin to address this deficit in the next chapter. We shall then go on to deal with the other obvious shortcoming so far, namely that we have not discussed how notions of time and sequence are to be addressed.

Chapter 3

Computation and entropy

If we wish to argue that a definition of computation in terms of self-informa-tion is minimally adequate, we had better work out how this encompasses a universal Turing machine. Intuitively, if we have a computable function, then we have no doubt that we can learn the unique value of the codomain for any value of the domain, given time and mental or physical effort. In other words, considering our domain and codomain as sample spaces of independent random variables X and Y respectively, if it is a function F intensionally encodable as $\mathscr{E}_i(F)$ in a random variable Z, as well as satisfying formulae (2.1) and (2.2), it satisfies

$$\forall d \in D. \; \forall c \in C. \; F(d,c) \Rightarrow$$
$$P\left(Y(c) \mid X(d) \cap Z\left(\mathscr{E}_{i\mathcal{L}_\mathbb{P}}(F)\right) \cap \mathcal{L}_\mathbb{P} \cap \mathbb{P}\right) = 1 \tag{3.1}$$

where $\mathbb{P}$ and $\mathcal{L}_\mathbb{P}$ (implicitly wrapped in random variables axiomatising physical uniformitarianism and epistemic limit of logic respectively) are the minimal background knowledge that will allow us formally to justify a belief in the effect of some computable function that satisfies F. Since they are constant and necessary only for considering proofs, we omit them from now on in this chapter for clarity. We note that the encoding of $\mathscr{E}_{i\mathcal{L}_\mathbb{P}}(F)$ in $\mathcal{L}_\mathbb{P}$ may be a predicate specification rather than an executable formula: the provision of a recipe for computing the function, provably so in an encoding $\mathcal{L}_\mathbb{P}$ of a logic $\mathbb{L}_\mathbb{P}$, is nevertheless the best way of *proving* that formula (3.1) is satisfied for a given F. This is important since if $\mathscr{E}_{i\mathcal{L}_\mathbb{P}}(F)$ had to be a *recipe* for the computation we would be giving a definition in terms of itself. We shall give an explanation of the relationship of the logic to this definition in Chapter 7, as well as how we avoid circularity in the definitions of $\mathbb{P}$ and $\mathbb{L}_\mathbb{P}$ themselves experimentally.

In formula (3.1) we are essentially replacing axioms of computability with an axiom about knowledge about physical reality, which itself is an abstraction about the consistency of the Universe in which we live across time and space, conditioned to be convincing by the evolution of our species and personal experience. If we think about the conditional probability, formula (3.1) is the same as saying that the probability of the intersection of $Y(c)$ and $X(d)$ is the same as the probability of $X(d)$. From the definition of conditional probability and rearranging, satisfaction of (3.1) means the same as satisfaction of

$$\forall d \in D.\ \forall c \in C.\ F(d,c) \Rightarrow$$
$$P(Y(c) \cap X(d) \cap Z(\mathscr{E}_i(F))) = P(X(d) \cap Z(\mathscr{E}_i(F))) \tag{3.2}$$

Computability corresponds to the practicability of constraining F in this way when given a physical encoding, or the security of deductive knowledge of c given d if F has truth-functional semantics. Where $\mathscr{E}_i(F)$ encodes the specification of a Turing machine, F is a member of the set of Turing-computable functions. We note that nothing in our constraints on C prevents its members from being infinite: if we were to have a Turing machine that computes π to ever increasing numbers of decimal places and never halts, it does not contain any more information than the algorithm that computes it.

Turning then to $P(Y(c))$, we need to consider the probability in the case where it is unconstrained, in order that we can characterise the information that it contains. While c might be unconstrained, C must be fixed for any given definition of F, as a consequence of formula (2.2). It is also possible, given (2.1) and (2.2), that F might also satisfy

$$\exists c \in C.\ \exists d, d' \in D.\ F(d,c) \wedge F(d',c) \wedge d \neq d' \tag{3.3}$$

in which case it is a surjection but not a bijection. If it does not also satisfy this, then it is a bijection. Since F is at least a surjection, we also know that

$$\forall d \in D.\ \forall c \in C.\ F(d,c) \Rightarrow P(X(d) \cap Z(\mathscr{E}_i(F)) \mid Y(c)) \leqslant 1 \qquad (3.4)$$

From now on we shall omit the first half of this implication for clarity, since it can be inferred by the presence of d, c and F.

From Bayes's theorem, we know

$$\begin{aligned}
&P(X(d) \cap Z(\mathscr{E}_i(F)) \mid Y(c)) \\
&= \frac{P(Y(c) \mid X(d) \cap Z(\mathscr{E}_i(F))) \cdot P(X(d) \cap Z(\mathscr{E}_i(F)))}{P(Y(c))}
\end{aligned} \qquad (3.5)$$

from which, together with (3.4), we can deduce through the independence of events and our rearranged computability definition of (3.2) that

$$P(Y(c)) \geqslant P(X(d) \cap Z(\mathscr{E}_i(F))) \qquad (3.6)$$

By exponentiating both sides of the definition of self-information[1] (with base b) of a random variable,

$$b^{-I_Y(c)} = P(Y(c)) \qquad (3.7)$$

and doing the same to the other side, taking $\log_b$ of both sides and rearranging (again using the independence of the random variables), we obtain

$$I_Y(c) \leqslant I_X(d) + I_Z(\mathscr{E}_i(F)) \qquad (3.8)$$

If F is a bijection and also reversible, there exists a function F' and corresponding Z' such that

$$I_X(d) \leqslant I_Y(c) + I_{Z'}(\mathscr{E}_i(F')) \qquad (3.9)$$

At first sight, a reversible function looks impossible thermodynamically

---

[1] As defined by Fano in [50, p. 34].

without losing information, until we realise that if the information carried in $Z(\mathscr{E}_i(F))$ and $Z'(\mathscr{E}_i(F'))$ is physically encoded such that it remains unchanged after the function (or its inverse) has executed, then it is present on both sides of the inequalities and can be cancelled out. If that configuration is reversible such that $I_Z(\mathscr{E}_i(F)) = I_{Z'}(\mathscr{E}_i(F'))$, then $I_X(d) = I_Y(c)$. Our axiomatisation therefore admits thermodynamically credible reversible functions. We know that a physical configuration can be constructed that encodes both $\mathscr{E}_i(F)$ and $\mathscr{E}_i(F')$, albeit not necessarily particularly efficiently at a first approximation.[2]

The usefulness of computing obscures the fact that the results of computations are merely showing information to us in a form that is useful because it elucidates some semantic relationship between the hypokeimenontic referents of the domain and codomain. Computations do not create any new information and are by definition pleonastic and vacuous. This also has a spatial analogue. Consider the 'devops' paradigm of creating a system image from a short machine specification: this may be less than a kilobyte in size, yet vast numbers of copies of the system *image* generated from it occupy gigabytes of hard disk space – still more if they sit upon a system that regularly snapshots them to multiple copies of back-ups.

In order to consider infinite codomains, we need the idea of coinductive types – and then cofixpoints[3] to generate the inhabitants of these codomains. Any prefix of a potentially infinite computable function's codomain contains no information as long as we define how much of it we want independently of the actual value of the codomain. This gives rise to an interesting interpretation of an abstraction of 'time' in the codomain: if any idea of sequence is axiomatised in the codomain, that sequence only gains information content 'when' it consumes information from a coinductive parameter. This provides a two-layered notion of time-like orderings – an internal ordering of states with the same information content in states between external

---

[2] The first report of this was by Bennett in [17], and since we are only discussing the *principle* here, we omit consideration of more recent literature.

[3] As formalised for example in the Coq proof assistant and whose history is summarised by Bertot in [18].

events and an external ordering of monotonically increasing information. However, the external ordering can only be perceptible internally if the function preserves information – in other words, it is reversible. If we try to fit physical reality to this model, we can deal with the reversibility of physics in the internal ordering, and the thermodynamic ordering of the arrow of time in the external ordering. This leads directly to a notion of time within the Cosmos that is the 'unfolding' of cosmontic information into an extensional time that can be axiomatised along with spatial dimensions as spacetime. The obvious relationship between such Shannon and thermodynamic entropy is Landauer's principle [90]. We shall return to this theme shortly.

If we want to add non-determinism to a computable function, we need to add entropy from somewhere else, but this is always equivalent to including the randomness as an entropy parameter in the domain – it has not magically appeared in the codomain.[4] We could run the function again with the same 'random' data and get the same result, as is the case if we run a pseudo-random number generator with the same seed. If we have a physically non-deterministic machine, we are stealing information from Nature, and must be dispersing some existing natural information in making the observation that encodes that extra information by Landauer's principle, or at least a qualitative version of it. Hence, if we consider the whole system, including physics, we are still obtaining the entropy from a parameter. We note as an aside that a non-deterministic Turing machine has no additional expressivity than an ordinary Turing machine, as one can always run through all the states of the non-deterministic machine deterministically (at the potential expense of time complexity, or accuracy if we sample one stochastically through a probabalistic machine). The same is true of quantum Turing machines such as those described by Deutsch [40], at the expense of yet more time.

The proof that a computable function's codomain contains no information is the function itself, treated through its inverse as a decompression function of the domain (we can always construct such an inverted version

---

[4] This is closely related to the 'entropic sandwich' discussion in [56].

of any function, even if inefficiently [17]). If there was any surprise in the value obtained then it would not be an effective procedure, as it would generate different values depending on when and where it was run. Through the Curry-Howard isomorphism, the determinism of computation can be thought of as the physical basis of the soundness of deduction in logical systems, and both essentially axiomatise the uniformitarian intuition that physics is invariant across time and space.

From these observations we intuit that extensional time as axiomatised by physics and intensional time as axiomatised in a sequence of pieces of information have a different quality, which we shall develop later. However, it suggests our first hypothesis directly:

Hypothesis 3.1. Intensional time is the experience of informational entropy unfolding extensionally into physical entropy.

Somehow, in the words of Lord Kelvin, '...it may I believe be demonstrated the work is *lost to man* irrecoverably; but not lost in the material world.'[5] We shall examine in Section 6.1 the nature of what exactly is lost in the production of Clausius's entropy [33, p. 357], enmeshed as it usually is in accounts of temperature and extrinsic state. It will turn out that thinking outside boxes of ideal gases is informative.

We note that examples of non-computable functions such as Radó's busy beaver [126] and determination of Chaitin's constant [30] involve asking questions about computable functions themselves in general *using* computable functions. It is hardly surprising that these types of function are uncomputable as their outputs would have surprisal value: they would be self-referentially trying to extract more information than is put in, falling foul of Gödel-type contradictions and also our definition of a computable function.

---

[5] From p. 5 of a text extracted from a preliminary draft of the *Dynamical Theory of Heat* (manuscript PA 128 of Kelvin papers at Cambridge University Library) at [139, p. 281].

Chapter 4

Causation and temporal specification

So far, we have reviewed what it means to have a complete specification and what it means for something to be computable. We have not yet said much about time, except to mention cofixpoints in the context of infinite codomains and the first intuition that this might have something to do with perception. Before we can fully relate the ideas of a structure of *extensional time* and a sequence of data marking *intensional time*, we need first a more basic extension to the axiomatisation of Section 2.3.

In order to carry this conception of specification as articulated in Section 2.3 into a temporal modality (for a specification F'), we must extend our domains with time and allow properties to be defined relative to sequences of states. We need to add extra restrictions to D and C when they are modal in time in order to axiomatise bounds on causation from the point of view of the reference frame. We shall base our restricted D and C on a common carrier set V, defining the instantaneous values of V as $W \triangleq V \times T$. T is taken to have the same structure in both domain and codomain, irrespective of whether D and C are disjoint. In the most general case T is a totally ordered set representing extensional time, which can be the set of real numbers representing a classical view of time. T can also be limited to discrete subsets of the real numbers. A partial order on W can be defined using the

23

well-ordering on T. If T is the set of real numbers greater than some given infimum, the total order is that of the arithmetic '$<$'. We use the operators $\underset{time}{<}$ and $\underset{time}{\le}$ to express strict and non-strict ordering relations respectively on W with respect to T. The use of the word 'time' here is abstract T time, and does not necessarily correspond to physical, wall-clock time.

It is implicit in the idea of 'a history' as opposed to a snapshot, that it is evaluated at some specific time up to which it has accumulated. If we are interested in the *most general* carrier set for histories, it would not make sense if we restricted the times at which values could be defined or the V could take for arbitrary T, so we need some equipment to restrict the powerset (set of all subsets) $\mathscr{P}(W)$ to make sure we have a consistent maximum T. In order to allow some flexibility for real numbers, we shall consider a supremum as opposed to a maximum, the difference being that a supremum is a least upper bound that does not necessarily belong to the set of interest. We therefore need some means, for any subset S of W, to refer to those elements of S that share a supremum when S is projected onto T, so we define a subsetting operator $\mathcal{S}_V$ using a suitable projection of S:

$$\text{proj}_T(S) \triangleq \{t \in T : \exists v.\ (v,t) \in S\} \tag{4.1}$$

$$\mathcal{S}_V(S) \triangleq \left\{x \in S : \exists v \in V.\ \exists t \in T.\ x = (v,t) \wedge t \le \sup\left(\text{proj}_T(S)\right)\right\} \tag{4.2}$$

This allows us to define a set of possible histories that all end at the same T. We can now give a first definition of the carrier set of histories,[1] which is the set of all histories of W excluding artificially sparse ones:

$$W' \triangleq \left\{ \begin{array}{c} S_S \subseteq \mathscr{P}(W) : \forall S \in S_S.\ \forall x \in W.\ \forall x_s \in \mathcal{S}_V(S). \\ x \underset{time}{\le} x_s \Rightarrow x \in S \end{array} \right\} \tag{4.3}$$

We notice that this allows histories with more that one V for each T.

---

[1] We use primed notations so that metavariables for subsets of powersets are not confused with the sets from which those powersets were constructed.

For engineering – though not necessarily for physics[2] – we are interested primarily in specifications implementable in classical physical models, so we can further restrict W' to rule that out and ensure each history has only one member of $V \times T$ for each T:

$$W'' \triangleq \left\{ \begin{array}{c} S_S \subseteq W' : \forall S \in S_S. \ \forall x \in S. \ \forall v, v' \in V. \ \forall t \in T. \\ x = (v, t) \land (v', t) \in S \Rightarrow v = v' \end{array} \right\} \tag{4.4}$$

If the domain $\mathcal{D}'$ and codomain $\mathcal{C}'$ of our classical temporal specification are subsets of W'', we can also specify that $\mathcal{D}' \cap \mathcal{C}' = \varnothing$ if we want to prohibit specifications from allowing the codomain to influence the future, which will be important in some of our later applications of general causal theories. We note that nothing in the causation relation confines the direct causative influence of an instant to the infinitesimally close following one. This is important in engineering where delay needs to be axiomatised, and in physics where quantum effects are determined by something that happened discontinuously in time. For engineering implementation or classical physics, the primitive element of causation is the most immediately preceding moment on the relevant number line. We can now axiomatise causality in zero spatial dimensions. First, we prohibit the admittance of redundant and causally impossible structure to F':

$$\forall D' \in \mathcal{D}''. \ \forall C' \in \mathcal{C}''. \ (D', C') \in F'.$$
$$\Rightarrow \left(\sup \left(\text{proj}_T (D') \right) \underset{time}{<} \sup \left(\text{proj}_T (C') \right) \right) \tag{4.5}$$

Next, we add the monotonic requirement, forbidding the editing of history:

$$\forall D' \in \mathcal{D}''. \ \forall C' \in \mathcal{C}''. \ (D', C') \in F'$$
$$\Rightarrow (\forall D'' \in \mathcal{D}''. \ \forall C'' \in \mathcal{C}''. \ D' \subset D'' \land (D'', C'') \in F' \Rightarrow C' \subset C'') \tag{4.6}$$

In words, this means that all functions of histories in the domain that are identical up until a given time must produce histories in the codomain that

---

[2] For example, we could consider Feynman path integrals.

are also identical up to the equivalent same time, irrespective of what happens in the future. In the physical case, any real theory is likely to restrict causation much more heavily, for example according to a light cone.

Considering infima, it would usually be the case that

$$\forall D' \in \mathcal{D}'. \; \forall C' \in \mathcal{C}'. \; (D',C') \in F'$$
$$\Rightarrow \left(\inf \left(\mathrm{proj}_T \, (D') \right) \underset{time}{<} \inf \left(\mathrm{proj}_T \, (C') \right) \right) \qquad (4.7)$$

but this is not necessarily so for specifications that start to elaborate a sequence of states in extensional time from constant initial conditions before being constrained by any input in intensional time.

Proof of equation (4.5) and equation (4.6) for some F' should proceed by induction and coinduction if we formalise a compliant F' using (at top level) a coinductive type that is dependent in[3] an explicit starting time, and which by construction contains an inductive predicate over its history.

In this chapter we have extended our axiomatisation of complete specification with time and causation, and have started to introduce the idea of intensional and extensional time. In the next chapter, we sharpen the distinction as we generalise the prescriptive-sounding 'specification' to 'theory,' add a minimal ontology and axiomatise some general properties of a topology suitable for descriptive, prescriptive and logical theories.

---

[3] This special use of the word 'dependent' takes its meaning from intuitionistic type theories, as discussed in Section 6.3: to paraphrase, it means that the type itself is affected by a value.

Chapter 5

Ontology of coordinate systems and theories

It is all very well having a set-theoretical notion of causation, but unless or until its structure is elaborated so that it can *model* something, it is of limited use. Even before we consider any spatial aspects, we need some way to refer to some*thing*, before we tie that thing to any concept analogous to physical 'world line,' be it continuous or discontinuous. In this chapter we shall be interested in two types of coordinate system. The first is a 3-tuple $\{I, V, E\}$ and the second is a 4-tuple $\{I', V', T', E'\}$ from which a history H is a member of $\mathscr{P}(I \times V \times E)$ and H' is a member of $\mathscr{P}(I' \times V' \times T' \times E')$. I and I' are sets of identifiers, V and V' are sets of values and T' is a set of extensional times. E and E' are sets of 'epistemic,' 'entropic' or 'event' intensional and extensional times respectively, and are constrained to be in one-to-one correspondence. E and E' map to the T of the causation constraints of Chapter 4. We shall come to the various roles of T' and other embeddings later.

We define an extensional causal theory $\mathcal{T}$ as one that satisfies our definitions of a computable function and causation, and also

$$\forall \mathcal{D} \subseteq \mathscr{P}(I \times V \times E) . \forall \mathcal{C} \subseteq \mathscr{P}(I' \times V' \times T' \times E').$$
$$\forall D \in \mathcal{D} . \forall C \in \mathcal{C} . \mathcal{T}(D, C) \tag{5.1}$$

A *preservative* extensional causal theory ('PECT') is one for which there also exists a $\mathcal{T}^{-1}$ such that

$$\forall \mathcal{D} \subseteq \mathscr{P}(I \times V \times E). \; \forall \mathcal{C} \subseteq \mathscr{P}(I' \times V' \times T' \times E').$$
$$\forall D \in \mathcal{D}. \; \forall C \in \mathcal{C}. \; \mathcal{T}(D,C) \Leftrightarrow \mathcal{T}^{-1}(C,D) \tag{5.2}$$

from which we deduce that we also want $\mathcal{T}^{-1}$ to be a computable function and thus $\mathcal{T}$ to be reversible. A theory that contains a parameter that cannot add information to a history is a theory with a parameter that cannot affect it, and it therefore has no business being a parameter. We thus largely confine our interest to those $\mathcal{T}$ that are also PECTS.

If $\mathcal{T}$ is a topology on $\mathcal{D} \times \mathcal{C}$ defining a topological space, then we define $\mathbb{T}$ as the set of topological spaces on $\mathcal{T}$ that are isomorphisms of each other, or homotopy-equivalent. Any Cartesian product on a $\mathcal{D}$ and $\mathcal{C}$ with a theory $\mathcal{T}$ determines a family of other such topological spaces that constitute members of $\mathbb{T}$.

We said in Chapter 4 that we can require of a temporal specification that $\mathcal{D}' \cap \mathcal{C}' = \varnothing$. In this chapter, we refine that idea such that C' can affect the *predicate* that admits monotone additions to D' (through an $I'V'T'$ hyperplane) but not the data so predicated. This corresponds to the intuition that an extensional theory may deterministically create a new variable, but once created, non-determinism may affect how the state of that variable evolves.

A theory $\mathbb{T}$ can be characterised as a cofixpoint that consumes a coinductive parameter from $\mathscr{P}(I \times V \times E)$ and determines a coinductive trace from $\mathscr{P}(I' \times V' \times T' \times E')$. Both domain and codomain are restricted by coinductive predicates. Such a cofixpoint must parameterise the domain predicate with its accumulated history at each step; this paradigm is well suited to axiomatisation in a dependent type theory. This concrete parameter to a predicate can be erased in any actual computation.

The difference between $\mathcal{T}$ and $\mathbb{T}$ can, where $\mathcal{T}$ is a logic, be correlated with the idea of *institutions* of Goguen and Burstall [61], which is foundational to the idea of heterogeneous specification as implemented by the HETS project [109]. Institutions provide a model-theoretic account of the equivalence of logical systems, abstracting any particular algebraic formalisation.

For a prescriptive $I'V'T'E'$ space in an engineering context, the larger it is, the less plausible it is to insist on a single formal reasoning system for a language axiomatisation or *ad hoc* constraints. As long as specifications are pairwise composable – that is, each coordinate in space S constrained by or referred to in a constraint has a structure morphism between all logics involved in the applicable constraint – then heterogeneous specifications will be consistent over S. While this is a practical argument for the concept of institutions where $\mathcal{T}$ determines $\mathbb{T}$, we shall also find in Section 5.1 that it is indispensable for inductive physical reasoning, where $\mathbb{T}$ determines $\mathcal{T}$. We shall further develop this in Section 6.3, in which we shall narrow the idea of institutions to require some minimal amount of expressivity before moving beyond them to topological foundations inspired by more recent formalisms.

In this chapter we shall first review the duality between completeness in engineering specification and physical theories in the context of our ontology, introducing special cases of $\mathbb{T}$, *viz.* $\mathbb{P}$ and $\mathbb{L}$, for physical and logical theories respectively. We shall then go on to consider the interpretation of $I'V'T'E'$ spaces in this context and conclude the discussion with a consideration of the topologies of PECT theories.

5.1 Completeness of theories over an *I'V'T'E'* ontology

In Chapter 2, we developed an idea of what it meant for a specification to be complete as the 'Goldilocks' middle ground between under- and over-specification. Over-specifying systems by programming implementations rather than an ontology we argue is the root cause of intractable obsolescence and technical debt. Our definition of a computable function that makes reference to logical and physical theories implies that there must be some canonical idea of each of these concepts if our definition of computability is to be properly tight. In order to avoid all these ideas becoming a piece of circular reasoning, we need to found at least one of them in something that is *in*ductive as opposed to *de*ductive. In Chapter 2, we critiqued verification and validation in engineering. We now pause to consider a similar question in physical theories. It is clear that a theory that cannot explain experimental results must be incomplete. This is how classical theories came to break down and quantum theory developed. In this case, classical physics is the inductive analogue of an underspecified piece of engineering. However, it is also the case that a physical theory that invents the existence of things that cannot be observed, tested, or which does not make any more accurate predictions than a smaller sum of leaner theories, is too flabby to be experimentally investigated, and is *over*specified: if it is not susceptible to the scientific method, then it is not scientific.[1] This leads to the question: what are the properties of an equivalent 'Goldilocks' physical theory $\mathcal{P}$? We suggest that it is a minimal theory that can put the history of the Cosmos in a $1:1$ correspondence with the minimum information required to determine the non-deterministic evolution of its primitive physical entities – in other words, a PECT theory in which the domain is the compressed information content of the Cosmos from the set of possible worlds that any candid extensional theory elaborates. A *concrete* theory $\mathcal{P}$ in $\mathbb{P}$ – as opposed to the abstract $\mathbb{P}$ itself – does require some particular logic $\mathcal{L}$. We shall tie this all together in Chapter 7 through axiomatising a mutual fixpoint on $\mathbb{L}$, $\mathbb{P}$, and the definition of a computable function. It does not matter if the information content of the domain is smaller than it appears because it turns out to be self-correlated to some extent outside a *physical* theory: any decompression function that can allow it to be inflated from a more succinct form can also transitively inflate the information in the codomain.

5.2   Interpretation of $I'V'T'E'$ spaces

An $I'V'T'E'$ space can be thought of as a stack of $I'V'T'$ hyperplanes for each successive member of E', where the extensional evolution of the $I'V'T'$ space is purely deterministic. If a physical theory is correct or a synthetic specification is realised correctly, E' unfolds at the same rate as T', and can be fixed in the $I'V'T'$ of the previous hyperplane at the point at which there

---

[1] This perhaps resonates with ideas about reality put forward by Whittaker (as digested by McConnell) [101, p. 58]: we arrive at not dissimilar conclusions by a different route.

is corresponding new information in the IVE space. New information at e in E defines a new $I'V'T'$ hyperplane at e' in E' that has a deterministic projection into the future of extensional time T'. It may be an infinite future – contingently waiting to be gainsaid for some event that may or may not happen – or it may be a finite future until some piece of information *must* enter through an increment within E. The hyperplane may be considered globally, such that nothing can evolve while waiting for a 'blocking' piece of new information, or it may be partitioned, such that some set of points belonging to a particular subset of I or I' can evolve independently, in which case both intensional and extensional time are partial orders when considered globally. The former describes engineered systems that are axiomatically fully synchronous or natural systems considered according to non-relativistic classical physics. The latter describes all other cases.

The semantic interpretation of E' differs according to whether we are considering natural or engineered systems, that is systems that are descriptive or prescriptive, embodying information that is cosmontic or anthropontic respectively.

In the case of an $I'V'T'E'$ space that describes a natural system, increments in extensional E' are accompanied by the appearance of new information that is a function of an increment in the E of the IVE space parameter. The hyperplane defined for constant E' then describes the evolution of the Cosmos according to the deterministic extensive time of a physical theory. Evolution of a point with a label in I' at a time in T' past the epistemic time in E' corresponding to a new piece of information emerging at the corresponding intensional time in E is counterfactual and thus physically irrelevant.

A synthetic specification over an IVE space that defines computable functions prescribes the creation of points in the corresponding $I'V'T'E'$ space for a given *synthetic* $I'V'E'$ hyperplane (remembering that the $I'V'$ can differ from the IV).[2] The synthetic IVE space aligns each epistemic

---

[2] We can perhaps regard this as axiomatising the observation of Babbage that 'At each stoppage every figure-wheel throughout the Engine, which is capable of being moved without breaking, may be moved on to any other digit. Yet after each of these apparent falsifications the engine will be found to make the next calculation with perfect truth.' [11, p. 67]

time in its E with the real extensional time in *natural* T' given the most recent relevant time in *natural* E', and axiomatises an anthropontic observation according to Definition 1.5. The $I'V'T'$ of this *synthetic* function are expected to be bisimulated by an implementation such that *another* anthropontic specification could observe a value in an $I'V'T'$ (correlated to a value in an $I'V'E'$) as a value in a *different IVE*. In other words, such a prescription specifies that an implementing system be engineered to make these points represented by a physical observable, which at some point it will fail to do.[3] It might produce evidence of a point that should never have been created or fail to produce evidence of one that should. Some faults of this type may be interpreted as correct by redundancy or a voting scheme in a weak bisimulation, but eventually some will go unmasked. Where does this leave the $I'V'T'$ ontology? The synthetic ontology is one *built* on observation – some points in an $I'V'T'$ space exist because of the fact or possibility of observation; others exist because they are defined to exist *given* the existence of other points. The *fallibility* of the observability of these points that are defined, deterministically, to exist, can be axiomatised by adding the possibility of error information in the IVE parameter at a lower level of abstraction; if that level of abstraction has an appropriate continuous or quasi-continuous[4] variable in its IVE or *is* a physical theory $\mathcal{P}$ in $\mathbb{P}$, this error information is that in any departure from an appropriate equivalence relation. We recall that a computable function is not supposed to produce new information: any such new information is a leak of non-determinism outside an equivalence class at a lower level of abstraction. If this possibility is included in the equivalence class of the computable function (*i.e.* it models the possibility of error), for an increment from e to e' in E' at $t_{e'}$ in T' – an $I'V'T'_{e'}$ hyperplane exists in

---

[3] If we are dealing with an engineering specification we require an equivalence relation between values anchored to E' or a T' branch that accommodates non-determinism arising from natural information gain and our partial ignorance of the *actual* starting state of the system. Otherwise, we would have to call everything an 'error': the maintenance of the equivalence relation under perturbation characterises one aspect of the stability of the system. It is also inherent in this that there are in theory infinitely many $I'V'T'$ hyperplanes for continuous models of time.

[4] *i.e.* its value is restated on *every* new information event.

the $I'V'T'E'$ space, which gives an ontology to how the $I'V'T'$ space defined at $t_{e'}$ evolves with time thereafter. The difference in the $I'V'T'_e$ and $I'V'T'_{e'}$ spaces axiomatises the error that occurs at $t_{e'}$ and ensures that a meaning can be given to how an implementation behaves after $t_{e'}$, even if it is 'incorrect' according to the contents of the $I'V'T'_e$ space for $t' \geqslant t_{e'}$. The relationship of an $I'V'T'E'$ space to the corresponding IVE space of a specification transformed to be tolerant of some set of errors in the E of some weakly bisimilar implementation is expressive enough to allow fault-tolerant properties to be stated and verified as a constraint on the congruent embedding up to a fault-tolerant interpretation of successive $I'V'T'$ hyperplanes given the noise explicitly introduced in E.[5]

5.3   Topologies over *I'V'T'E'* spaces

The deterministic subject matter (*i.e.*, for constant e') of a PECT theory in the codomain is at first sight a subspace $S_{L'}$ in which the identifiers I' are constrained to be members of a finite set of labels L', $L' \subseteq I'$, of cardinality N. This can be thought of as a set of N canonical coordinates v'_i, $0 \leqslant i < N$ on a suitable phase space. This in turn suggests that a computable function could be formulated as a Hamiltonian, as Penrose has suggested [118, p. 231]. While this might be the case for a practical computable function with some proven bound to its time and space utilisation (*i.e.* primitive recursive), it cannot easily represent a Turing-complete model of computation because this would require a phase space in which evolution of state can influence the dimensional cardinality of the space within which it moves. This implies that the kinds of topologies we are interested in may have fragments that can be modelled as symplectic manifolds, but in general something richer is required. Once we relax the constancy of e', the non-deterministic elements left to be determined by a parameter would be incompatible with Liouville's theorem. A PECT theory may therefore be statically parameterised over some

---

[5] This is a development of a concept first introduced in [56].

initial subspace S_L but the actual size of the space may expand dynamically,[6] in a manner not dissimilar to the π-calculus [104, 105].

The types of theories that we shall discuss in the next chapter can all be viewed as non-trivial topologies, and this will be important in our efforts to integrate them in Chapter 7. This makes it likely that a formalised fully integrated theory should be couched in homotopy type theory [155] rather than the naïve – or indeed zF [159] – set theory we have, for simplicity, adopted in this exposition.

---

[6] We shall see later that given the Cosmos seems to have a finite amount of matter and energy, there is an interpretation under which Liouville's theorem may be applied, but it only seems to work at the scale of the entire Universe.

Chapter 6

Axiomatisation of theories

In this chapter we shall examine the characteristics of three types of complete theory – physical,[1] prescriptive and logical. The epistemology of these inclines to an empiricist, scepticist and rationalist perspective respectively, and we shall attempt to integrate these aspects in Chapter 7. The nature of logical theories is induced by physical theories, and the nature of prescriptive – or *anthropontic* – theories is induced by both, and fills out the theory initiated by our analysis of engineering specification in Chapter 2. We shall assemble these parts fully in Chapter 7, but first we need to characterise them individually. We shall start with physical theories, as these condition the extent of the other types. We are aware that this resonates strongly with ideas of, *inter alia*, Parmenides, Aquinas and Victorian natural theologians and theistic evolutionists. While we do not think the ideas expressed here are inconsistent with the metaphysical views of Babbage or Lord Kelvin among others, a scientific tract such as this is not the *forum conveniens* for that sort of enquiry. '*Hypotheses non fingo*' [114, p. 530], to borrow some famous words of Newton. We shall not therefore pursue the question of where the information in the parameter of a PECT theory comes from here, save to investigate what the scientific method, which implicitly axiomatises uniformitarianism, can, and *cannot* tell us about it.

---

[1] A subtype of descriptive.

6.1 Physical theories

It seems plausible that a solution to Hilbert's sixth problem [70, pp. 454–455]
exists – that there is a theory that is *minimally* adequate to explain the extensional time evolution of the Cosmos, given some information that satisfies
the requirement for computability. However, naïvely, it seems possible that
it could be an infinite theory – or to be more precise – have an information
content greater than or equal to that of the Cosmos. However, this would be
a universe with no inductively perceptible physical laws, and the success of
the scientific method in characterising the Universe sufficiently deterministically to make accurate predictions and do things such as build computers
tells us this is not so. Enough of the structure of $\mathbb{P}$ – call it $\mathbb{P}^-$ – is already
known[2] to construct mechanistic proof checkers for rich $\mathbb{P}^-$-consistent logics
$\mathcal{L}_{\mathbb{P}^-}$ in $\mathbb{L}$. This fact, together with the compactness theorem, strongly suggests that $I\,(\mathbb{P}) \ll i_{max}$ (where i_{max} is the maximum information content of
the Cosmos), which is *almost* an existence proof. If the extensional Universe
suffers a heat death, our interpretation of time would imply that an underlying computable function characterisation would terminate, consume no
more information and *epistemic, intensional* time would end at its asymp-
totic limit, even if extensive extensional time would continue, Narnia-like,
as part of the fabric of a system in a deterministic adiabatic equilibrium. If
the only *events* in the Universe happen in intensional time, and we adopt a
Lamport-like view of time [88], we have a strong hint that *physical perception
is a function of intensional* time. If new information is intensively discrete,
then maximal entropy and thus closest approach to absolute zero temperature has some prospect of being reached. If it were to turn out that there is no
heat death and $i_{max} = \aleph_0$ (the cardinality of the natural numbers), this style
of attempt at an existence proof of the sixth problem would fail, although a
constructive proof by supplying a compliant theory would not. The debates
about the ultimate fate of the Universe can hardly be regarded as settled, so
let us play it safe and assume we are going to need to find a constructive
proof. The primacy of causality in a PECT theory suggests that the theory

---

[2] Although to be pernickety, including over-specified aspects, we might call it $\mathbb{P}^{\pm}$.

we are looking for might have a similar shape to theories of causal sets [21], with partial orders playing an important part. However, we would ideally prefer some new physics to relate quantum mechanics and general relativity, so we intuit that we need to arrive at a causal theory by a completely different route. We would certainly like to avoid a mathematical retread of an existing theory, or there is no reason to believe one theory of quantum gravity over any of the others – and we are not aware that any of the proposals made so far really answers Hilbert. In a spirit of curiosity, let us explore some different ideas.

We have established in Section 5.1 that PECTness is a reasonable acceptance criterion for a physical theory, but there are a number of fundamental problems in any search for a PECT that embodies the ($\mathbb{P}^-$) Standard Model of physics. First, quantum field theory appears to suggest that the vacuum carries far more information in its perpetual perturbations than is substantiated by macroscopic evidence. Second, general relativity has no mechanism for gaining information. Dark energy casts a pall over both problems, both as an account of vacuum energy, which must be finite given its gravitational effect, and in that for the purposes of general relativity it appears as a phlogisticated fudge factor with no causal explanation and no consenus as to its energy content. Is this 'ganglion of irreconcilable antagonisms'[3] actually irreconcilable, or can some topsy-turvy time paradoxes actually lead to a solution?

Let us start by considering what Nature allows us to do with information. There seem to be two fundamental possibilities: first, information can be encoded or observed; second, in a controlled system, it can apparently be lost into uncontrolled degrees of freedom when work is done on it. Both can be related to energy. Let us consider the encoding case first.

The obvious candidate to relate information content with the carrying capacity of energy is the Whittaker-Shannon sampling theorem [137, 158], from which we have the Nyquist frequency $B = f_s/2$ (where B is the bit rate and f_s is the sample frequency) – making energy and mass equivalent to

---

[3] This delightful piece of verbiage is from the libretto of *H.M.S. Pinafore* by W. S. Gilbert [59, p. 21].

Shannon entropy. Using the Planck-Einstein relation $E = h\nu$, we can obtain the amount of information encodable per SI units of energy and time as $\frac{1}{2h}$ ShJ^{-1}s^{-1}. If we consider the units of $E = h\nu$ in the case where E is in Sh and h is 1 or other dimensionless number, then ν, which in an extensional manifestation would have units of Hz, should have an intensional dimension of Sh, which implies *extensional* time has units equivalent to Sh^{-1}. This implies an interpretation in which time in the wave equation is an extensional measure of how spread out the information it contains is along the spatial propagation dimension, measuring a time *relative* to the intensional 'tick' that supplies the next event. We can change to any system of extensional units we choose by manipulating the units of h. It is also possible to view the energy of particles relative to other particles – kinetic or potential energy of various forms – as the potential to encode information. The fraction of the information mass of a particle to its encoding ability would be determined by h. One could give energy different units to Shannons[4] to capture the scale difference, if there turned out to be one. We note also at this stage that by this time-as-spread-out-information line of reasoning, we must have time dilation as in special relativity [46] just by the Doppler effect [43] in the manner of Bondi's k-factor [23] and conservation of energy: this is consistent with but does not require Maxwell's equations [98].

Taking our cue from Maxwell, let us test the idea with some dimensional analysis. In the absence of a dimensional symbol for information,[5] we shall use the base 2 unit of Shannons. If time has dimensions of Sh^{-1}, we might guess that there is a time-state symmetry implied by the Schrödinger equation [132], and length in space would have to have dimensions of Sh$^{-1/3}$. If energy has dimensions of ML^2T^{-2} and units of Shannons, then we deduce that mass must have units of Sh$^{-1/3}$, *via* M $\left(\text{Sh}^{-1/3}\right)^2 \left(\text{Sh}^{-1}\right)^{-2}$ = Sh. Similarly, velocity in this scheme therefore assumes units of Sh$^{-1/3}\left(\text{Sh}^{-1}\right)^{-1}$ = Sh$^{2/3}$. This must be wrong, since our argument about special relativity implies that c must be a dimensionless unity in order to conserve information and energy.

---

[4] The Shannon is defined by the International System of Quantities IEC 80 000-13 [75]; ISO/IEC 80 000 is a conceptual dependency of the SI system [25].

[5] We might perhaps use E for ἐλευθερία, for reasons that will become apparent.

If we are to raise Shannons to a zero power for our velocity units, we must choose spacetime as the symmetry (with Sh^{-1} for length and Sh^{-4} as the dimensions of spacetime). Mass now has intensional units of Shannons: the same units as energy. We can now return to the question of kinetic energy – or more generally, energy implicit in position and momentum, which leads us to the next point.

Nature allows us to do work with information, but hotter systems seem to require more work to encode information in them *artificially*. Landauer's principle, which sets out to quantify the work potential lost when a bit of physically encoded information is dispersed in a thermodynamically irreversible process, considers a two-state system using Boltzmann's entropy formula to derive $E_L = k_B T \ln 2$, where T is the temperature of the environment [90]. The principle has not been found inconsistent with experiment (*e.g.* [19, 80]), but there are three main difficulties with it arising from the extensive nature of Boltzmann's entropy formula when looking for a causal link between the microstate and the macrostate: the first concerns temperature; the second concerns what it means to have *any* kind of extensive state when observation of it is fundamentally limited by Heisenberg's uncertainty principle [69]; and the third arises from the implicit causality inversion present if we over-extend what can be inferred from a phenomenological coarse graining argument. We shall try to deal with these problems as we edge towards a possible causative account of what is going on. Let us first consider temperature.

A first attempt at defining temperature usually states that it is a quantity that is proportional to the average kinetic energy of some set of particles. However, this soon starts to go wrong. Naïvely, we might imagine that something will get 'hotter' proportionally to the heat we put into it, but this is not what happens, as materials have differing heat capacities; worse, those heat capacities vary depending on how much thermal energy the material already has; worse still, some of this energy is not even stored as straightforward kinetic energy, but in condensed matter especially (rather than a monatomic ideal gas, which is usually the starting point for an explanation), there are all kinds of potential and kinetic modes in which the energy might be stored, only some of which seem to contribute to the temperature 'felt' by neighbouring bodies in a naïve, *translational* kinetic energy sense. The whole idea of 'translational' kinetic energy only provides a reasonable explanation in the case of gases or plasma, in which individual particles are not coupled for longer timescales than it takes them to interact and bounce apart. It ignores all of the thermal modes of a solid black body radiator that give rise to the characteristic emission spectrum at a given temperature, irrespective of that body's composition.

We can see then why we need some concept beyond energy to explain how 'hot' something is. Usually, the next stop in this sort of discussion is the idea that temperature is a relative concept defined in terms of the phenomenology of heat transfer: something is hotter than something else if it tends to give up some of its energy to its counterpart until the two bodies are in a state of thermal equilibrium, in which they are defined to have the same temperature. This is fine as far as it goes, but it is qualitative only, is uncomfortably teleological, and does not provide an account of absolute temperature, which is indispensable when considering what happens to isolated black body radiators.

These sorts of considerations lead, when combined with the insight that the further two bodies are from thermal equilibrium the more scope there is to extract useful work, to the idea of entropy as coined by Clausius. Entropy is a conserved quantity in a Carnot engine, in which heat and temperature together are reduced to a quotient. However, this does not tell us what temperature *is*. The intractability of thermodynamic temperature usually leads to the definition

$$T = \frac{1}{\left(\dfrac{\partial S}{\partial U}\right)_{V,N}} \tag{6.1}$$

i.e., the reciprocal of the rate of change of entropy with respect to internal energy for constant volume and number of particles. However, this becomes problematical on close approach to absolute zero and does not tell us what *entropy* actually is, unless we already know what temperature is, which we do not.[6] Usually in physics, if we cannot attach a physical meaning to an important mathematical quantity, either we lack physical understanding, or the mathematical theory is to some extent wrong; entropy as a quotient is too powerfully predictive to be *very* wrong. Breaking the circularity of entropy usually requires an appeal to an increase in the number of microstates with increasing entropy or some nebulous idea of 'disorder,' but the former is not a causal explanation and the latter is neither quantitative nor very informative about the nature of the 'order' that it negates. Neither is it clear how or why from first principles the number of microstates would acquire units of JK^{-1}, albeit multiplied by an arbitrary constant. In 2019, the sɪ Kelvin was redefined using the Boltzmann constant as the quotient relating it to energy in Joules [25, p. 127]. This is a matter of practical metrology and does not help with characterising a hypokeimenontic quality of temperature that is convincing hyp*er*keimenontically. The previous definition of Kelvin was '...1/273.16 of the thermodynamic temperature of the triple point of water' [26, p. 114], with zero on the Celsius scale offset to 273.15 K by the minute difference between the ice point and triple point of water.[7] The situation becomes even more complicated when considering industrial *measurement* standards. Leaving aside this historical tangle, it is slightly uncomfortable that Boltzmann's constant has units partly in terms of a quantity that it is now supposed to *define* – in the case of the metre defined in terms of the speed of light such an approach seems reasonable, because we have a good account of what length *is*, but the same cannot really be said about temperature, which remains stubbornly elusive. The units of entropy strongly imply that temperature is just a dimensionless ratio relating two amounts of energy, which is *not* the amount of heat stored in a material but a quantification of the availability of that heat, in some rather hand-waving way. Didactic – as opposed to Socratic – progress in physics usually requires that these sorts of awkward questions be quelled, which is a pity.

This conundrum leads us, through considering the Carnot cycle [28], to

---

[6] Feynman observed that we do not even know what *energy* is [51, vol. 1, no. 4, p. 2], so *a fortiori* it is hardly surprising that temperature is in even worse shape.

[7] Water itself is defined in relation to reference quantities of different isotopes in H_2O.

an alternative hypothesis:

Hypothesis 6.1. Extensive thermodynamic entropy is the available extensive self-information of a system encoded in its state in a reference frame containing only the particles constituting that system.

Which suggests another:

Hypothesis 6.2. Work is the transfer of extensive information encoding capacity without an accompanying transfer of uncontrolled information.

Which together suggest:

Hypothesis 6.3. Temperature is the ratio of (a) the difference between the extensive information encoding capacity of a body's thermal energy in a reference frame containing only the particles constituting that system and the self-information available to be encoded to (b) that same self-information.

Or

$$\frac{U - S}{S} \tag{6.2}$$

We cannot just have U in the numerator of this quantity, or for $S > U$ we would have a quantity describing a system with more entropy than it could encode, which is offensive to the holographic principle and thus unacceptable for reasons we shall develop later. We do not have a dimensional problem in our expression, as both U and S have dimensions of information under this interpretation. For this to work and not produce negative or infinite temperatures, entropy can never be zero or exceed the information encoding capacity of a body. Likewise, absolute zero can never be reached, which is uncontroversial.[8] We shall consider the consequences of this further in due course. Unlike equation (6.1), Hypothesis 6.3 is a proper function of state rather than a property of a hypothetical change or transfer of heat. The only apparent problem is that it is not immediately clear why a body burdened with more information would give up its heat more slowly.

---

[8] This is because temperature is an extensive, classical property of real-numbered physics. We shall have more to say about intensive interpretations and the maximum entropy of the Cosmos later, and the one condition under which $T = 0$ makes sense.

We shall address this presently. Nevertheless, with an information hypothesis, we at least now have a statement of what thermodynamic entropy *is*, which – if it is true – is an improvement: the circularity is gone.

Given that this is a somewhat radical suggestion, we shall pause briefly to consider an alternative view. If entropy were not the primary quantity with units of energy (or information), then, by $k_B T$, temperature must be. Temperature defined purely in terms of energy would lead to a very cumbersome measure when we are interested in the temperatures of various and usually inhomogeneous substances in the practical business of zeroth law comparisons. This is because, stripped of any concept of rate, the only relationship available is with the internal thermal energy of the body whose temperature we are interested in. There appears to be a bijective function for any substance from heat to temperature that is an intrinsic property of that substance, but it is a horrible function, and would bake the properties of water into the interpretation of temperature in Joules in an invisible and hence more troublesome way than they are baked into Kelvin. The Debye model of solids [37] – of which more shortly – can improve the situation, but is too approximate for many purposes. If we persist with the nasty bijection, and elide the redefinition of thermodynamic temperature in 2019 and of Celsius in terms of Kelvin before that, pretending Celsius and Centigrade are the same thing, then the thermal energy that a Kelvin represents in 1 kg of H_2O is implicit in the difference of the isochoric heat capacities of liquid water at 0°C and 100°C, where Centigrade is defined relative to freezing and boiling points at a fixed pressure. 1 K *would* be a $1/100^{th}$ of this amount of energy, but the amount of heat required to bring about a given increase in the proportion of the energy that can be 'felt' is not linear, specific heat embodying *total* thermal energy, which is itself a function either of temperature or internal thermal energy. The non-linearity of the relationship would make Planck's law [123] impossible to write down without the temperature-heat function of water, which is worse than just having the constant k_B. This would obscure the molar amount of the substance of interest, and cause problems with the quantitative statement of Landauer's principle.

We shall now examine Hypotheses 6.1 to 6.3 in more detail. We have said that extensive thermodynamic entropy is the extensive self-information of a system available to be encoded, but this suggests three questions: what do we mean by 'extensive' entropy as opposed to any other kind; what do we mean by 'self-information' in this context; and what does it mean to be available to be encoded without again becoming mired in teleology – is there information that is *not* available to be encoded, and if so what is it? First, by extensive entropy, we mean that we are talking about the normal entropy that is a function of state and which can be measured *in some reference frame*. It does not carry any history with it save that which could be deduced by knowing the precise history of the transitive closure of all the particles that previously interacted with it. Second, by self-information, we mean the actual extensive information content in this history, not as it might be duplicated, obfuscated or amplified in some deterministic way: we suggest that much confusion between thermodynamic and information entropy occurs because thermodynamic entropy and informational entropy are *not* the same, thermodynamic entropy being instead *self-information* as defined by Fano [50, p. 42]. Third, by 'available to be encoded' we mean the information as convoluted by its various interactions with external systems. It is *not* sufficient to characterise the precise, real-numbered kinematic parameters of every relevant particle, but is the maximum *relative* information that can be recovered in the reference frame of the system. Its availability to be encoded is not teleologically contingent on its actually being encoded, but rather is a statement of the relative information content of some particular subset of particles. The total of all systems in the Cosmos (arbitrarily and disjointly partitioned) together would contain all of the historical information, but any one of them would have some mixed up function of some subset of information through the application of Liouville's theorem at a system-of-systems level that is unknown at the level of a particular subsystem. It is a partitioned subset of this mixed up information that is available to be encoded, and we suggest that it corresponds to the concept of the information encoded in the extensive microstates of the system of interest.

The information definition of work of Hypothesis 6.2 suggests that work is a transfer of extensive information coding capacity without accompanying information: it is therefore available to encode something else. Considering a Carnot cycle, during adiabatic expansion, the entropy in the cylinder does not change. Therefore, according to our idea of extensive entropy, the information available to be encoded cannot have changed either. What *has* changed is the amount of entropy that is received along with the energy – a cooler body will transmit more entropy with each unit of energy. The fidelity with which the information in that entropy will be encoded if it is allowed to cool spontaneously will be greater for a body that has been heated by work done on it. However, a smaller quantity of unique self-information will be transmitted per quantum of energy. A hotter body will effectively encode the same information more than once, increasing what could be regarded as a signal-to-noise ratio if we were interested in capturing all of the available information. We might therefore expect that if we do work to a body to raise its temperature, we could amplify that information at the expense of whatever information gleaning capability was lost in bringing the work to bear. This brings us to the informational definition of temperature.

According to this line of thinking, the informational temperature of a body is a function of state as a ratio of quantities of information that are also functions of state. To say that temperature is a tendency of a process to happen is a phenomenological explanation without aetiology and thus not really very suitable as an extensive account of what is going on, let alone an *intensive* one. However, any explanation based on state must be *compatible* with a phenomenological account. To this extent we can also think of temperature as being a measure of how fast that body's extensive entropy is lost relative to the rate of information flow from it (if it is flowing kinetically to a cooler body or being radiated as photons). These rates should converge as thermal equilibrium is approached asymptotically. For a black body radiator, the informational temperature correlates with the instantaneous *power* available to drive thermal electromagnetic emission, which according to the Stefan-Boltzmann law should relate to the actual power in Watts in si units.

Hypothesis 6.2 seems reasonably convincing from the point of view of a Carnot cycle. However, we would like to have at least a qualitative causative mechanism. If we consider the Debye model of a solid, we can think of phonons as encoding information, with an encouraging symmetry with the Whittaker-Shannon sampling theorem. An idea of thermodynamic temperature then emerges from the average excess energy carried by a phonon in comparison to the energy of a photon with an equivalent information content. There is a plausible link here, with the more energetic modes at higher temperatures leading to the emission of more energetic photons with lower entropy content. We shall need an indirect way of quantifying the link *via* a conservation law since analytic phonon models make simplifying assumptions. However, before examining this further, we must first look at a potential wrinkle in even the idea of a conservation principle.

In these considerations about extensive entropy and temperature, we have been thinking about the Carnot cycle and the special case of the Clausius inequality [33, p. 219] in which no new entropy is created. We have only been considering it moving around. We note that as it moves, the information recoverable tends to spread out and become convoluted or gains *mixed-up-ness*, to use Gibbs's word [58, vol. 1, p. 418]. However, we know that this increase in the convolution of information and decrease in the practicability of disentangling it does not in itself imply an increase in *total* entropy as a state function. This has historically led to much confusion, not least with the supposed Gibbs's paradox. Jaynes [78] points out that the paradox is not real, citing Gibbs's *Equilibrium of heterogeneous substances* [58, vol. 1, pp. 55–353]; however, nothing in the non-existence of the paradox addresses our problems with causality inversion. This hints that something else is also happening here and the two aspects of the second law – equilibration and entropy increase – ought to be better separated. Where does new information come from, and what has this to do with temperature? To make progress with this, we need to think more carefully about what is going on in a black body radiator.

As few discussions about physics are complete without a sphere in a vacuum, let us invoke a sphere of some solid material of a size so as to give it some amount of internal kinetic energy in its comoving frame and leave it infinitely far from anything else for eternity: eventually, it will approach the state in which all of its thermal energy has been radiated away as photons in accordance with the theory of black body radiation. At a quantum level, some of the thermal interactions of electrons will have caused excitations that led to non-deterministic electron energy level changes. As the thermal energy tends towards zero, nearly all capacity to encode information in the thermal motion has been lost extensively from the system. If the entropy were lost too, this would seem to accord with the holographic principle. Yet there must be some extensive entropy somewhere at absolute zero; the third law of thermodynamics tells us only that its rate of change tends to zero. Where has it gone? Somehow a set of particles that in some reference frame is at absolute zero has no information locally available to be encoded, but globally it cannot have vanished. Perhaps it is all now in thermal photons? But what about the configuration of the sphere itself? One conceptual mistake here seems to be to have assumed that it is meaningful to think that at maximum entropy absolute zero of extensive temperature has been reached 'at infinity.' We shall return to this presently, but let us first step back a little.

We have just skipped over a problem with our unexciting black body sphere in that all the while the latent extensive information in the thermal movement was going on, assuming the assumptions of the adiabatic theorem [24] cannot be met, quantum energy level changes will have occurred to some extent 'resulting' from quantum field perturbations *inside* our sphere. Perhaps we should look at this more closely. There is non-determinism here, which requires information to drive a PECT theory – but where has it gone if it is to be preserved? And where – physically – has it come from for that matter? Thinking about the motion of particles, by our previous argument about convolution, the new information would be combined with information in thermal movement in such a way as it would be irrecoverable in the extensive Universe, even in principle by some kind of hypothetical Laplace's daemon.[9] If our definition of temperature in formula (6.2) is to survive, we would have to assume either that somehow extensive entropy has increased, and therefore heat capacity is not an absolute intensive property of a substance, which seems peculiar, or we must violate the holographic

---

[9] Laplace assumes complete determinism *ab initio*, but that mistake does not make the concept *entirely* uninteresting [91, p. 4] (Laplace does not use the word 'daemon').

principle, which in our model is quite serious, since if information and energy are the same thing, breaking it is tantamount to contradicting the first law of thermodynamics.[10] We could decide that the holographic principle requires that the new information generated inside the isolated sphere was never more real than a tree that falls silently in the forest. This idea might be sustainable on a precarious extrapolation of Penrose's Cosmic censorship hypothesis [117], but it jars with Einstein's intuition, and with no means to test any prediction, fails to meet the basic desiderata of a PECT theory. Variation of heat capacity with time cannot be correct by a sort of proof by contradiction: at some point temperature could turn *negative*, which is meaningless. We cannot fix this just by adding more energy from nowhere, as this would also break the first law of thermodynamics; doing so by *any* route lacks credibility and seems an unpromising feature of any prospective account of reality.

There seem few ways left to resolve this paradox, which is troubling, since a closely related implicit axiom to uniformitarianism in science is that Nature seems to abhor a contradiction. We could abandon the idea of information as energy. Or we could look to a possibility of an *intensive* cousin of extensively encoded information for an account of the *apparent* darkening burden of new information on energy. If potential information is an intensive property of mass or energy with an attractive gravitational effect, it seems reasonable to ask whether actual information would have the opposite effect. Let us consider whether the consequences of this would be believable or helpful. First, we would seem to have extinguished some information encoding capacity – to us, it has gone 'dark.' We seem to have dark energy with a repulsive gravitational effect. The matter associated with the new information has become *lighter*. That seems like an idea that could solve a few problems, although it might seem appropriate to elevate the significance of the Landau–Lifshitz pseudotensor [89, p. 306]. What about our *extensive* information paradox? If something has become lighter, it must

---

[10] The first law is also consistent with the idea of a PECT theory taking *actual* information as a parameter that is *given*; the *primum movens* or original genesis of that information is a *meta*physical question.

curve spacetime less, a change that should propagate through spacetime, conveying information in the extensive movement of other matter and energy and complying with the holographic principle – but how would this wave be initiated? We notice that there is another consequence of something suddenly becoming lighter – it would lead to a loss of gravitational potential energy, and we would be back in trouble with the first law. However, if space abruptly expanded through a spatial dilation sufficiently to counteract the effect, this would certainly lead to a gravitational pulse (which could be characterised as a graviton), and one that would have an effect that tended to become isotropic throughout space as extensive time tended to infinity. Each individual expansion event in a patter of 'little bangs' would be extremely small; the Big Bang in such a theory might be equally thought of as the 'First Bang': all are singularities insofar as these interactions can be considered points with before-and-after real-numbered extensive positions in spacetime. If this is true, we shall see that other singularity problems seem to disappear. This model would give us an expanding Universe without a need for cosmological constants, de Sitter spaces or vacuum energy pressure. When we think about this carefully, it is not clear how much space should expand, as it would depend on other particles: it is more credible to think of energy in an expansive microgravitational wave propagating at the speed of light, of an initial amplitude determined perhaps (supposing a spatially closed universe) by its asymptotically isotropic limit. It is puzzling how *any* global curvature superimposed on spacetime from outside could be reconciled with the first law (without vacuum energy from nowhere) unless it was this *curvature* that caused effective changes in mass: it is as perplexing as considering gravitational potential energy as something that is lost from infinity because it follows nicely from integrating Newton's law of gravitation with respect to distance, even though we now know nothing started out that way. We shall develop our alternative explanation further, consider more of the cosmological implications of this theory, assess whether it is compatible with observation, and address what sort of experiments might support or undermine such a theory shortly, but first we return to the sphere in the vacuum to clear up some points about Landauer's principle.

If we reprise the sphere but now imagine it with a surface covered in dimples like a golf ball, it would have a finite number of spatial wells, which would also be tiny gravitational potential wells or electrical potential wells if the sphere had a net charge.[11] We could fill or not fill these dimples with particles to encode something like a conventional two-dimensional barcode, but covering a closed 2-manifold. What would be the energy cost of performing this encoding? If the golf ball was at zero-point energy, along with the particles in the wells, we could perhaps move them. Suppose Maxwell's daemon[12] has grown hot and tired of operating hatches in boxes and has picked up some tweezers. Suppose too that he and the tweezers have also had time to cool down to zero-point energy. If the daemon moved the tweezers very slowly, he could pick up a particle and move it to another well in an adiabatic limit as his speed tended to zero. No *natural* extensive entropy has been created, but apparently 'information work' has been done. We are not in trouble with $E_L = k_B T \ln 2$ because we have rigged T to be zero. Now, more – if not fully – realistically, let us imagine that the golf ball is at 1 K but the tweezers are still at absolute zero. Let us also assume – again unrealistically – that the tweezers have not absorbed any energy radiated from the golf ball. What is the energy cost now of doing work on our bit of information? There are two extremes. At the first, the tweezers can simply grab the particle by jabbing in the right sort of place, and given the right combination of scales, with a chance of success perhaps comparable with swatting a fly. The particle-tweezer ensemble, if left for an infinitely long time, would radiate the kinetic energy that the particle had enjoyed. The tweezers could now move as in the absolute zero case and adiabatically place the particle in a neighbouring well. At the other extreme – even less realistically – perhaps

---

[11] Potential wells have been used to explore Landauer's principle experimentally in [19, 80].

[12] The 'daemon' is spelt by Lord Kelvin without the 'a' ([147], p. 326); Maxwell does not use this expression at all [99, pp. 308–9]. It is notable for our later discussion in Chapter 7 that Kelvin uses the expression 'free-will' (*loc. cit.*); Maxwell uses words and phrases like 'artificial' in [99, p. 151] or 'unaided by any external agency' (*ibid.* p. 153'); Maxwell also uses the phrase – attributed to Kelvin – 'inanimate material agency' (*loc. cit.* and for original remark [146, p. 265]) in the context of the impossibility of deriving work from heat reservoirs arranged in the contrary direction to that of a Carnot engine.

the daemon could align the tweezers perfectly in position and velocity with the particle. However, this is essentially the same as the Maxwell's daemon hatch scenario, and fails by the same sort of argument as a Szilárd engine [142] as a putative method of frustrating the impossibility of perpetual motion machines: whatever the smallest particle imaginable, some quantity of energy would be dissipated in the information to tweezers or hatch feedback loop. However, perhaps the tweezers have a mechanism to store kinetic energy of the thing they grab in some kind of reservoir, maybe some idealised spring, in which case the whole process could potentially be made theoretically adiabatic *without* any kinematic knowledge, if we pretend that all of this can be done without any side effect on the wider environment.

The conclusion we reach in this argument is that the work done given naïve considerations is purely that in arresting the particle in all its degrees of freedom. The hotter the system, the more work that must be done. This is qualitatively consistent with $E_L = k_B T \ln 2$, which gives us the amount of energy lost to non-information bearing degrees of freedom scaled with temperature. However, given we could always, as in an MRI scanner, change the state of matter without arresting a particle in a matter-matter interaction,[13] simplistic considerations of the manipulation with apparently arbitrary configurable parameters of matter – in both molar measures and mass – do not provide an explanation: the quantitative aspect of Landauer's principle seems difficult to prove at the smallest scales. Moreover, there is no firm cut-off between the classical regime in which particles stay where they are put, and the quantum scale at which they can jump about randomly, and with a probability that can be enhanced by even the slightest amount of thermal energy, which is ultimately fatal to this kind of attempt to conceptualise a microstate: in the quantum limit it is inadequate as either an intensive or extensive explanation of what is happening.

If we are going to preserve Landauer's principle, we shall need a deeper account of information manipulation than we can access by arguing about gross manipulations of matter or energy, which is just a form of arbitrary

---

[13] Vaccaro and Barnett have investigated a reservoir of angular momentum rather than energy [156].

coarse graining. We can do this by forgetting about manipulating the spatial or electromagnetic state of matter in the mechanical or electronic engineering paradigm and instead imagine we were encoding information in
the thermal excitation modes of a Debye solid. While not entirely realis-
tic, it is easier to see how the energy ratio of a self-information-equivalent
phonon and photon would give us the work done in having allowed one bit
to be lost to uncontrolled degrees of freedom and in injecting a replacement
according to our whim.[14] In a classical limit, 'microstates' would seem to
be real-numbered, and the information content would look fractional when
there is less than 1 Sh. The Landauer limit energy at a given temperature T
corresponds to the binary limit of such a system. We have a logarithm to
base 2 and $S = 1$, so by formula (6.2), $E_L = T + 1$. We thus quantitatively re-
cover a slightly modified version of Landauer's principle in which as T tends
to absolute zero, E_L tends to unity. This reduces to the standard definition
of Landauer's principle for $T \gg 1$, modulo some constant factor incorpo-
rating a change of base and k_B in Sh. We thus have a choice: we can either
have a hard quantitative view of Landauer's principle for an encoding in the
temporal (or spatial) frequency domain in which we systematically neglect
quantum effects or settle for a qualitative Landauer's principle in the Sh^{-1}
extensive time domain. However, we notice that the only plausible mechanism of ultimate *physical* agency (through tweezers or otherwise) in the second option once it has entered Nature – we assume diabatically through the
quantum mechanics of our central nervous systems – is through the first.[15]
To a large extent this is all angels dancing on the head of a pin because the
Landauer limit in practice corresponds to an extremely small amount of energy and practical problems likely supervene before it is reached.

Returning to our main argument, let us sum up our progress so far. Consideration of the Carnot cycle and black body radiators has led us to the

---

[14] Our definition of work (by the sphere or external process) is met here because the information is not uncontrolled: work is always implicitly defined relative to some anthropogeni-
cally framed teleology, whether or not involving direct or indirect human agency.

[15] This can be thought of as an extrapolation of Penrose's ideas about the simulatability of
minds [118] or Hofstadter's 'strange loop' [72]. We shall return to this subject later.

idea of a smaller gravitational interaction for matter with a more eventful history. Later in this section we shall consider how this nexus of matter and history may play out in a full theory. First, though, if information is to be so central to a theory, it is time that we considered how a standard system of units might relate to the unit of information, the most basic of which, by ISO/IEC 80 000, is the Shannon, which we have already discussed in our dimensional analysis. This might give us a first hint as to whether the rate of the gravitational information content associated with matter is low compared to its mass, as it would have to be, given the apparent constancy of Newton's gravitational constant G.

The most natural dimension to start with in an information conversion is time: we ought to be able to relate energy and time by converting to Shannons through dimensional analysis. We note there are, by the SI definition, 9,192,631,770 oscillations of the unperturbed ground state hyperfine structure transition of the ^{133}Cs atom emission photon per second [25, p. 127]. By the Nyquist frequency, if it takes two oscillations to encode 1 Sh of information:

$$1\,\text{Sh} = \frac{9\,192\,631\,770\,\text{Hz}}{2} = 4\,596\,315\,885\,\text{s}^{-1} \tag{6.3}$$

i.e.

$$1\,\text{s} = 4\,596\,315\,885\,\text{Sh}^{-1} \tag{6.4}$$

The Planck-Einstein relation or the dimensions of the Planck constant h [25, p. 127] suggest our next move. The quotient JSh^{-1}, taking energy to *be* information encoding potential and therefore $h = 1$, can be obtained thus:[16]

$$1 = \frac{h \cdot \Delta \nu_{\text{Cs}}}{2}$$
$$= 6.626\,070\,15 \times 10^{-34}\,\text{J} \cdot 4\,596\,315\,885\,\text{Sh}^{-1} \tag{6.5}$$
$$1\,\text{Sh} \approx 3.045\,551\,149 \times 10^{-24}\,\text{J}$$

---

[16] These are rough calculations so we shall not be too hung up on the nicities of significant figures and error estimations.

JSh^{-1} is effectively dimensionless, as both are measures of energy in this interpretation, so we could multiply through by a ShJ^{-1} convenient conversion factor if we wish to convert conventional units to Shannons.

Looking now at distance, we can obtain a conversion to Sh^{-1} in terms of metres by means of normalising c [25, p. 127] to unity in Shannon spacetime:

$$299\,792\,458\,\mathrm{ms}^{-1} = \frac{1\,\mathrm{Sh}^{-1}}{1\,\mathrm{Sh}^{-1}} \tag{6.6}$$

$$\frac{299\,792\,458\,\mathrm{m}}{4\,596\,315\,885\,\mathrm{Sh}^{-1}} = 1 \tag{6.7}$$

$$1\,\mathrm{m} = \frac{4\,596\,315\,885}{299\,792\,458}\,\mathrm{Sh}^{-1} \\ \approx 15.331\,659\,49\,\mathrm{Sh}^{-1} \tag{6.8}$$

$$1\,\mathrm{Sh} = \frac{4\,596\,315\,885}{299\,792\,458}\,\mathrm{m}^{-1} \tag{6.9}$$

We can now say something about distance and time. What about temperature? If we turn again to the Boltzmann constant and consider whether a quantity in JK^{-1} can be converted to something that does not involve Kelvin, we need to look for something that relates energy to temperature. Neglecting the difference in Celsius and Centigrade, we could try and use the fact that $1\,\mathrm{K}$ is $1/100^{\mathrm{th}}$ of some non-linear mapping of the change in thermal energy in $1\,\mathrm{kg}$ of H_2O in the difference of the isochoric heat capacities in the liquid state at $0°C$ and $100°C$; but we have observed that this is useless, because we are interested in a ratio of heat available as a consequence of its temperature to total heat and we cannot derive the non-linear mapping in any obvious way from a single scalar number. Absolute properties of water are therefore irrelevant to our problem, doubly so since the redefinition of the SI Kelvin. However, recalling our definition of temperature of expression (6.2), in which it is given as a dimensionless ratio, Boltzmann's constant can be given units of Shannons. We shall normalise this to unity for the time being, though we shall come to question this decision shortly.

We thus obtain a conversion for Kelvin through the Boltzmann constant [25, p. 127]:

$$1\,\mathrm{Sh} \stackrel{?}{=} \frac{2 \cdot 1.380\,649 \times 10^{-23}\,\mathrm{JK}^{-1}}{h \cdot \Delta\nu_{Cs}}$$

$$\stackrel{?}{=} \frac{1.380\,649 \times 10^{-23}\,\mathrm{JK}^{-1}}{6.626\,070\,15 \times 10^{-34}\,\mathrm{J} \cdot 4\,596\,315\,885\,\mathrm{Sh}^{-1}} \tag{6.10}$$

$$1 \stackrel{?}{\approx} 4.533\,310\,107\,\mathrm{K}^{-1}$$

$$1\,\mathrm{K} \stackrel{?}{\approx} 4.533\,310\,107$$

As a matter of mathematical taste, it would seem from this argument that we can either have a Boltzmann constant of $1\,\mathrm{Sh}$ and temperature as a dimensionless ratio or dispense with k_B as dimensionless unity and give temperature units of Shannons. However, expression (6.2) implies that it must be a dimensionless quotient in order to maintain dimensional consistency.

We now pause to consider the Stefan-Boltzmann constant:

$$\sigma = \frac{2\pi^5 k_B^4}{15c^2h^3} \approx 5.670\,374\,419 \times 10^{-8}\,\mathrm{Js}^{-1}\mathrm{m}^{-2}\mathrm{K}^{-4} \tag{6.11}$$

If our set of units is self-consistent, we would expect that putting h, k_B and c into a calculation for σ should yield the same as using SI units and performing a unit conversion. If $k_B = 1\,\mathrm{Sh}$ while c and h reduce to dimensionless unity, then we might want to check whether:

$$\sigma \stackrel{?}{=} \frac{2\pi^5}{15} \cdot \mathrm{Sh}^4 \tag{6.12}$$

We plug in our SI constants for time, distance and temperature:

$$\frac{2\pi^5}{15} \cdot \frac{k_B^4}{c^2h^3} = \frac{2\pi^5}{15} \cdot \frac{\left(\frac{k_B}{\mathrm{JK}^{-1}}\,\mathrm{JK}^{-1}\right)^4}{\left(\frac{c}{\mathrm{ms}^{-1}}\,\mathrm{ms}^{-1}\right)^2 \left(\frac{h}{\mathrm{JHz}^{-1}}\,\mathrm{JHz}^{-1}\right)^3} \tag{6.13}$$

so we want to know:

$$\frac{\left(\frac{k_B}{JK^{-1}}\right)^4 JK^{-4}s^{-1}m^{-2}}{\left(\frac{c}{ms^{-1}}\right)^2 \left(\frac{h}{JHz^{-1}}\right)^3} \overset{?}{=} Sh^4 \tag{6.14}$$

Substituting and simplifying:

$$\frac{\left(\frac{k_B}{JK^{-1}}\right)^4 \frac{Sh}{\frac{\Delta\nu_{Cs}}{2s} \cdot \frac{h}{JHz^{-1}}} K^{-4}s^{-1}m^{-2}}{\left(\frac{c}{ms^{-1}}\right)^2 \left(\frac{h}{JHz^{-1}}\right)^3}$$

$$= \frac{\left(\frac{k_B}{JK^{-1}}\right)^4 \frac{1}{\left(\frac{\Delta\nu_{Cs}}{2s}\right)} Sh \cdot s^{-1}m^{-2}}{\left(\frac{c}{ms^{-1}}\right)^2 \left(\frac{h}{JHz^{-1}}\right)^4} \cdot \left(\frac{\frac{k_B}{JK^{-1}}}{\frac{\Delta\nu_{Cs}}{2s} \cdot \frac{h}{JHz^{-1}}}\right)^{-4}$$

$$= \frac{\frac{1}{\left(\frac{\Delta\nu_{Cs}}{2s}\right)} Sh \cdot s^{-1}m^{-2}}{\left(\frac{c}{ms^{-1}}\right)^2} \cdot \left(\frac{1}{\frac{\Delta\nu_{Cs}}{2s}}\right)^{-4} \tag{6.15}$$

$$= \frac{\left(\frac{\Delta\nu_{Cs}}{2s}\right)^3 Sh \cdot \left(\frac{\Delta\nu_{Cs}}{2s} Sh^{-1}\right)^{-1}}{\left(\frac{c}{ms^{-1}}\right)^2} \cdot \frac{1}{\left(\left(\frac{\Delta\nu_{Cs}}{2s}\right) Sh^{-1} \cdot \left(\frac{c}{ms^{-1}}\right)\right)^2}$$

$$= Sh^4$$

So far, so unremarkable. However, the interesting conclusion from this exercise is that on an information interpretation, starting from the knowledge that we are looking for a law quantifying energy per area per time (*i.e.* Sh^4), we can apply the reasoning in reverse to arrive – modulo $2\pi^5/15$ – at the Stefan-Boltzmann law by dimensional analysis and Shannon conversion alone. The fourth power of temperature is necessary to reach the correct dimensions in Shannons given the other constraints: we can introduce a c^2 to eliminate the length dimensions, followed by an h^3 to eliminate time, which requires a fourth power of energy through Boltzmann's constant in order to balance the Shannon powers in the equation. The presence of a π^5, albeit derived through Bose-Einstein statistics, is slightly suggestive of a relationship to the proton-electron mass ratio μ, which, it has been observed [94], is close to $6\pi^5$. This invites further investigation, but there is a more pressing issue: the 'spare' factor must be reconciled with our axioms about temperature, energy and information. Let us 'solve' this for now by revisiting equation (6.10) with a k_B that accommodates this phonon-photon scale relationship:

$$\sqrt[4]{\frac{15}{2\pi^5}}\, \mathrm{Sh} = \frac{2 \cdot 1.380\,649 \times 10^{-23}\,\mathrm{JK}^{-1}}{h \cdot \Delta\nu_{Cs}}$$

$$1\,\mathrm{Sh} = \frac{1.380\,649 \times 10^{-23}\,\mathrm{JK}^{-1}}{6.626\,070\,15 \times 10^{-34}\,\mathrm{J} \cdot 4\,596\,315\,885\,\mathrm{Sh}^{-1}} \cdot \sqrt[4]{\frac{2\pi^5}{15}} \quad (6.16)$$

$$1\,\mathrm{K} \approx 11.457\,436\,36$$

Assuming the relationship between inertial mass and information to be constant, we can complete our main four unit conversions by converting 1 kg to Shannons *via* $E = mc^2$, setting h to be unity:

$$E_{1\,\mathrm{kg}} = \frac{1\,\mathrm{kg} \cdot \left(299\,792\,458\,\mathrm{ms}^{-2}\right)^2}{6.626\,070\,15 \times 10^{-34}\,\mathrm{J} \cdot 4\,596\,315\,885\,\mathrm{Sh}^{-1}} \quad (6.17)$$

$$\approx 2.951\,029\,505 \times 10^{40}\,\mathrm{Sh}$$

so

$$1\,\mathrm{kg} \approx 2.951\,029\,505 \times 10^{40}\,\mathrm{Sh} \quad (6.18)$$

We can see by inspection that dimensional analysis would give the same result. We also note that this is a very large number: losing a comparatively small amount of information encoding ability may make a difference on Cosmic timescales, but perhaps not so on human ones.

We now have conversions for the dimensions of M, L, T and Θ. This allows us to convert all other units (the sɪ base dimensions of I, N and J can be trivially derived from the main four). Unlike the Planck natural units [124, pp. 479–480], we do not have an obvious way to characterise the gravitational constant G. Let us try and convert the Einstein gravitational constant [47] by doing a simple unit conversion to fill this hole:

$$\kappa \approx 2.077 \times 10^{-43} \, \text{N}^{-1}$$

$$\approx 2.077 \times 10^{-43} \left(\text{kg} \cdot \text{ms}^{-2} \right)^{-1}$$

$$\approx \frac{2.077 \times 10^{-43} \left(4\,596\,315\,885\,\text{Sh}^{-1} \right)^2}{\dfrac{1\,\text{kg} \cdot \left(299\,792\,458\,\text{ms}^{-2} \right)^2}{6.626\,070\,15 \times 10^{-34}\,\text{J} \cdot 4\,596\,315\,885\,\text{Sh}^{-1}} \cdot \dfrac{4\,596\,315\,885}{299\,792\,458}\,\text{Sh}^{-1}} \qquad (6.19)$$

$$\approx \frac{2.077 \times 10^{-43} \cdot 6.626\,070\,15 \times 10^{-34} \cdot 4\,596\,315\,885^2 \cdot \text{Sh}^{-2}}{299\,792\,458}$$

$$\approx 9.698 \times 10^{-66}\,\text{Sh}^{-2}$$

Like the mass of the Universe (about 10^{100} or $\left(10^{10}\right)^{10}$ Shannons if we wave a finger in the air and assume there is the equivalent of about 10^{60} kg of matter in the Cosmos),[17] this is a constant that we cannot manipulate to unity. We shall revisit the issue of why neither κ nor G (which is only a factor 8π different in our units) nor indeed such things as the mass of basic particles might be very 'nice' numbers shortly, in the tradition of Dirac's musings on the mutability of constants [41]: for this kind of investigation to make progress, we would need to explain the possible variations of some ratios but not others.

If this theory is to go forward, we need some convincing way to account for how an interaction history might attach to matter, and why quantum field theory seems to be correct – apart from its predictions about zero-point energy that are wildy inconsistent with the apparent current value of a cosmological 'constant.' Perhaps quantum fluctuations of the vacuum are extensionally necessary only insofar as they have causal consequences: maybe quantum field theory is just the mathematics that allows the probability – and therefore information content – of some interactions to be quantified

---

[17] Ten to the power of a hundred is sometimes called a 'googol.'

in terms of their predecessors. Under that interpretation, a vacuum really is a vacuum, free of mysterious bubblings: that does not mean that perturbative *phenomena* are not real – they have measurable effects so obviously are – but we suggest that a given bubbling is not *physically* real unless it is connected in a way we shall come to presently. The expansive pressure, we might infer, is an expansion in space owing to the dispersion of quanta of microgravitational waves. Meanwhile, a particle – or nexus between history and matter described extensively by a world line – curves and is affected by the curvature of spacetime to a degree that is determined by some metric on *all causal history*. Expansive propagation of gravitons would explain why expansion is observed to be isotropic even though matter and material interactions are not. This would suggest that dark energy could be thought of as an intensive dual to the self-information of thermodynamic entropy: a kind of Cosmic recording medium to which we have no read access, of the sort which in anthropogenic affairs is of interest to auditors.

We are beginning to think in terms of a division between the extensional, extensive, real-numbered, deterministic evolution of systems, and the information 'creating' events that occur in quantum interactions. The *real-numbered* parallel superposition of causative influences as they evolve in extensive time to give the extensive information content of an interaction could be thought of as a quantum computation. If nothing is actually localised anywhere unless and until a diabatic interaction happens, the apparent paradoxes of double slit-type experiments and quantum entanglement pose no problems; this also accords with the idea that *physical perception* is somehow linked to events rather than continuously evolving states. However, we have an embarrassing problem in our theory: if information somehow sticks to matter, what happens to the information in interactions of ancestor particles? How would it be allocated to descendants?

The theory for thinking about quantum interactions of particles is to be found in quantum electrodynamics (QED), originated by Dirac [42] and developed intensively by many over the middle decades of the twentieth century. Instead of attempting to give a potted history of QED and its various extensions, we shall focus on Feynman's diagrammatic conceptualisation of what is happening, noting and eliding extensions to the electroweak and strong fields and conventional issues of field and dynamical theory unification. Rather than proceed from the field theories, we shall attack the issue from a direction inspired by structures familiar in computer science, postponing the details of how the two approaches meet in the middle: this will lead us shortly to suggest a generalisation framework that might be required to explain stupendously high energy particles in the early Universe that might exist at far higher energies than current quantum field theories. We shall confine our discussion here to QED without extensions in the interests of clarity.

Feynman diagrams can be considered as directed acyclic graphs for some set of particle interactions, in which particles as edges join interactions as vertices, although the semantics 'give' the vertices to the electrons. The particles can be seen as representing transitive causation between interactions, where a diabatic interaction that takes place is a stochastic process requiring new information, if non-determinism is a parameter to a PECT theory characterisation. The less likely the interaction, the more information implicit in its having occurred. However, there is no reason for there to be a distinction between the aleatoric and epistemic if we take an intensional, Bayesian view – we can be agnostic about whether what appears as aleatoric surprisal is devoid of hyperkeimenontic meaning (*i.e.* truly random), or like the output of a good pseudorandom number generator, simply inpenetrable without the epistemic key. We can also view a Feynman diagram as a Bayesian belief network if we label each vertex with the information value *given* its antecedent events and the deterministic, real-numbered, statistical evolution of particle positions in the meantime. If we follow this to its logical conclusion, the entire history of the Cosmos could be viewed as a join-semilattice in which a Big Bang primaeval particle is the least upper bound, recovering something very like Lemaître's somewhat ignored primaeval atom or 'initial quantum' concept [93], which we find extraordinarily prescient.

If explicit non-local hidden variables are allowed (the explicit non-locality avoids any trouble with Bell's theorem [16]), then perhaps every particle for gravitational purposes is linked to the information (surprisal) in every antecedent particle, with its information entropy spread between *all* its descendants, weighted according to the extensive information carrying capacity (or mass) ascribed by its nexus with matter. The information in this type of causal information matter – perhaps call it *patter* – could be allowed to have the needed *repulsive* gravitational effect, or equivalently, reduce the *supra*relativistic mass of particles that cause curvature in spacetime.[18] In the terminology of Chapter 5, the topology of the lattice at any E' effectively encodes the label space, while all extensional intensive and extensive quantities are embedded in $V' \times T'$. Any massless gauge boson would seem not to *have* patter active in the continuous gravitational theory but act as a *vector* of patter.

This is encouraging, but not quite enough. QED tells us about interactions, but not about bound particles. We also have the outstanding problem that we need a schema or metatheory to accommodate the various QED-type theories through some dual dynamical theory to quantum field theory unification. These problems can have a common solution if we introduce some elements of cyclic heterogeneity to our directed acyclic graph. If we allow there to to be arbitrary *non*-directed connections, representing bound particles in the overall directed acyclic structure, we can give the graph a hierarchical interpretation: at the top level, it is directed and acyclic, and within the transitive closure of a non-directed island there may be directed acyclic content or hierarchical structure. What would be the inductive base case of such a structure? It must be the primaeval particle itself, whose energy we could never replicate, owing to the physical extension of the self-reference paradox. If we were to learn about such particles, it would have to be by induction through some kind of fractal-like structure of particles and subparticles, with *clues* – not, we emphasise, explanations – in such relations as the Koide formula [85]. If this were true, the Higgs boson would seem likely to be the very smallest in such a series. There is no logical limit on the scale of hierarchical structures except that the particles are bound: it therefore has the potential to be a homogeneous description of particle physics and

---

[18] We shall refer to *nexi* where we need to consider *in*tensive properties, and reserve 'particle' to refer to their dust-like *ex*tensive manifestation.

chemistry, which would make the wave-particle duality of large molecules less perplexing. In a non-directed structure, patter would equalise, which would mean that quantities such as the proton-electron mass ratio μ would be constant and the same as the ratio of their absolute, non-suprarelativistic masses. This is consistent with contemporary observation [12], but is not proved by it. There seem to be three types of possible vertices:

1. Bosons

2. Fermions

3. Interactions

Bosons and fermions may be composite, represented by a hierarchical graph, which could be encoded in a hypergraph, albeit with more visual clutter. Interactions may be either *admissive* or *dismissive*. An admissive interaction is one in which there is information in the fact that the quantised arrangement of a static configuration has changed; a dismissive one, one in which there is information in the fact that it has not. We have covered the vertices in our graph. What about the edges? There seem to be four types:

1. Acyclic interaction consequence connection

2. Fermion interaction association (may be transitive if required by the relevant field theory)

3. Acyclic extensive time evolution

4. Binding connection (may be transitive and cyclic)

Some example fragments of such a graph are shown in Figure 6.1, which introduces by example this schema of *suprarelativistic nexus dynamics*. The interactions shown are intended to show abstract topology and do not reflect the actual physics of any particular interaction according to the relevant field theory – but the nexi drive the resonances, rather than the conventional opposite interpretation. The information content of an interaction is shared between the transitive closure of non-directed connections and transmitted through directed connections in proportion to the mass of the peer or successor nexi respectively. The patter of a given nexus is therefore a metric on the lattice. The topology of historical predecessor nexi can be thought of as placing those particles 'in' those particles extant now: to that extent, the primaeval particle could be thought of as being 'in' everything that is in the Universe. Similarly, any two things that share a diabatic interaction in their history – which is any two things that have interacted in a remotely interesting way – share some patter according to a sort of ladder of interactions ascending to the hypothesised primaeval particle.

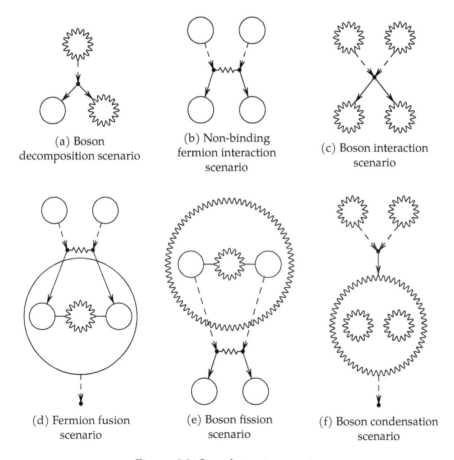

(a) Boson decomposition scenario

(b) Non-binding fermion interaction scenario

(c) Boson interaction scenario

(d) Fermion fusion scenario

(e) Boson fission scenario

(f) Boson condensation scenario

Figure 6.1: SND abstract scenarios

In Figure 6.1, bosons and fermions are represented as sinusoidal and plain circles respectively, as an *hommage* to Feynman diagrams. We adopt the computer science convention of drawing trees upside down. Dots represent interactions, which have a position in time and space in continuous, real numbered extensive coordinates. Solid lines with arrows represent transitions that occur in zero elapsed extensive time. They represent an instantaneous jump to a new configuration. Dotted lines represent the vector between a particle configuration arising from an interaction at one space and time to another, in the spirit of Feynman's simplified explanation of QED [52, p. 85]. Until the next interaction actually occurs, a localised probability distribution of causally feasible particle propagations evolves with extensive time, and this probability is related to the information content of the next interaction. Zig-zag lines represent interactions of fermions that retain their identity, but equalise patter between themselves. Boson-boson interactions must be singular. Interaction lines can enter a hierarchical fermion or boson (*e.g.* Figure (6.1d) or (6.1e)), but not a boson condensate (*e.g.* Figure (6.1f)). Figures (6.1d) or (6.1e) could equally have swapped the enclosing particle types. Bosons and fermions may only be joined by non-directed solid lines for heterogeneous connections: a fermion can only be connected to another fermion *via* a 'force exchange' boson. Figure 6.1 is shown as planar graph fragments, but the global graph need not be planar. Interactions between particles inside different hierarchical particles that do not change their extensive configuration would seem to create a braid between constituents of the knots of the fermions or bosons, as if exchanging an anyon. For that matter, the entire lattice could in fact be viewed as a braid of divergent and convergent strands. This schema likely requires refinement in a tight axiomatisation, but it seems at least a reasonable *ansatz*.

It is axiomatic that there is only one lattice for a descriptive PECT theory that we experience as the physical world. If this axiom is consistent with Nature, it would imply that there are other bosons that disintegrated in the early Universe and conceivably fermions that were confined long before the quarks. In any event, if this theory schema is correct, there is a great deal of new physics to be done here; some of the particles might be accessible in colliders yet to be built. It seems plausible that some chain of events that produced a large quantity of patter in a short space of time could account for a particularly rapid phase of Cosmic inflation.

The allocation of 'patter' to matter would result in the matter having a *liberature*, defined as

$$\frac{M - P}{P} \tag{6.20}$$

where M and P are the quantities in Shannons (or information in a logarithm of some other base) of matter and patter, or potential and actual intensive information. They are the intensive cousins of extensive U and S respectively in Formula (6.2), but relate energy implicit in matter as 'darkened' by patter rather than extensive energy rendered unusable (*i.e.* unable to encode anything else) by extensive entropy. This gives us an equation at maximum entropy, and in which the reference frame for our definition of extensive entropy of Hypothesis 6.1 is the entire spatial extent of the Cosmos:

$$\frac{U_{max} - S_{max}}{S_{max}} = \frac{M_{min} - P_{max}}{P_{max}} \tag{6.21}$$

would be the case at absolute zero of temperature and liberature, at which new information would cease to enter Nature and nothing would have freedom to exercise any agency. The extensional and intensional sides of $\mathbb{P}$ would be related through $P = i$ and therefore $P_{max} = i_{max}$. Knowledge of S_{max} is that which would be attributable to Laplace's daemon at the end of time, if the daemon existed.

If the extensive entropy S and intensive entropy P were equal (we shall revisit this hypothesis shortly), it implies that at heat death, so would U and M be: a half-and-half split of thermal energy in various forms with non-interacting massive particles saturated with patter and unable to interact further, which with no inertial mass, could travel at the speed of light – not so much weakly interacting massive particles (WIMPS), but not interacting at all. If things could move at the speed of light and did so, they would seem not to experience any extensive time. Whether this cold, static, Universe would eventually settle into some kind of exotic condensate in a fixed size Cosmic resonant cavity is an interesting question. If it did, the lattice could have a single meet infimum at the base of a complete lattice, whose nature seems like something that could have been imagined by Dante.

If, *extensively*, temperature equilibration is just a regression to the mean, a similar phenomenon would occur with liberature, except that no liberature would be transferred during an ordinary elastic interaction. Any island of intense interaction that did not interact strongly with other matter would tend to accumulate a local disequilibrium of liberature, which would reduce the net gravitational effect of its associated matter on the curvature of space-time. The cosmological implications of this would be quite interesting. In particular, the tendency to generate patter would operate as a gravitationally limitative process in very dense, very hot, very strongly interacting matter, which would suggest that according to such a theory, black holes might avoid unmanageable singularities and instead resemble something more akin to Chapline's theory of dark-energy stars [31], and we would not need to invoke Hawking radiation to predict or explain their evaporation. They would also tend to evaporate much more quickly than otherwise thought. The more closely packed matter became, the more possibilities there would be for interactions per unit mass, and the more information would subsist in whichever eventualities did transpire. Patter could therefore avoid a complete singularity inside black holes with information constantly radiated as gravitational waves, maintaining the holographic principle irrespective of what might be happening near the event horizon – but what about other cosmological problems? Where might dark matter fit in? Why is there more matter than anti-matter? Also, for a nascent theory with topology front and centre, we had also better be able to say something about the global geometry of the Universe, and perhaps suggest why the gravitational constant does not appear to be very special. It would also help if we could find some evidence for any of this. In order to make some progress, we need to revisit what might have happened in the very early stages of the Universe if the patter theory were to be correct.

We have developed the idea that intensional time is determined by 'ticks' of information flowing as a coinductive stream. What would happen at intensional time $t = 0$? Let us suppose that the Universe could have begun with the information content of the splitting of a primaeval particle. That particle would have associated with it the patter with an information content necessary to describe the way in which the first particle split. The first particle, however, is in a curious situation. First, we note that if there was one primaeval particle, then it would have to be *either* matter or antimatter, which would break the symmetry. That is helpful; the next implication is trickier. It has no gravitational potential energy at the time at which it splits because that time is the *beginning* of extensive time when (intensionally) there *is* no space (or *ex*tensional 'when' for that matter); the primaeval particle must intensionally *precede* both. However, there is another form of energy that might be in play – the kinetic energy of the products of the first split: but where is the space into which they are to move? If space had an open topology, then, compelled by Archytas's spear,[19] it might just expand wherever the particles went, but this does not offer an adequate account of what it would mean for space itself to expand. If on the other hand, space is a closed 3-manifold, why would it expand from a one-particle state, and how do we know what sort of closed 3-manifold it would be?

To the first question, there seem to be two possible answers: either we can invoke the first law again, and try to apply 'Nature abhors a contradiction' to mandate the creation of some space for the particles to have somewhere to move, or we can hypothesise that the way that massive particles acquire their kinetic energy – apart from any repulsive non-gravitational field in play – is from an expansive kick from the impulse in continuous time of an expansion event in *dis*continuous extensive time, giving us two gravitation-related processes driving expansion, but no messy hybrid gravitational and kinetic expansion regime. In an awkward mixed theory, we could not be entirely sure how big the Universe would have to be before dropping its kinetic expansion driver: if we wanted to allow time for further splits before a premature collapse it would depend on the exact energies, decay times and

---

[19] The argument originates with Archytas but the spear idea comes from Lucretius [96, ll. 968–983, pp. 33–34].

decay products of the particles in question, be hostage to non-determinism and seem causally questionable. We notice that with an expansive gravitational wave arising from both mass conversion and patter accumulation, the information load suggests a reflection of liberature in graviton quasiparticles as an analogous dual to the temperature of a phonon.

Thinking about this further, we might ask what happens to a particle that already has kinetic energy and becomes lighter – would it just speed up to preserve *kinetic* energy? There are two ways we can think about this. First, in the resultant comoving frame of an interaction,[20] energy may be preserved by speed increase in one or some of the interaction successors, insofar as they have resultant mass. Second, if time sped up owing to a time expansion relative to some other frame, this might help us, although only transitorily: we shall return to look at this more closely in a moment. First though, we shall consider the other limit of these particles – when they are totally saturated with patter and there can be no *kinetic* energy. These fast-moving zombie particles would only be able to 'reappear' in an interaction with a particle with greater liberature, and there would be nothing stopping them moving at the speed of light. Once everything had a liberature of zero, no further interaction would be possible, with photons having no suprarelativistic or *available* energy in much the same way that massive particles would have no suprarelativistic or effective mass. We begin to wonder if there would really be much difference between a gravitational wave travelling at the speed of light and one of these inert particles of no suprarelativistic mass doing likewise. Light, matter and gravity start to look very similar, and wave-particle duality again does not seem so very strange. Imagining what would happen to a particle that had *almost completely* lost all of its suprarelativistic inertial mass, we have a hint from the 'missing' energy in our k_B in Shannons that some kinetic energy may be finding its way into microgravitational waves, of which some may impart a kinetic kick to neighbouring matter. This seems to be about as far as we can go with this without a fuller theory, but it is a fruitful thought experiment.

---

[20] This is the only frame that makes sense, as there is no absolute frame in which to measure this energy unless we consider the entire Cosmos, which violates local causation.

Moving now to the nature of a closed 3-manifold, can we deduce what sort of closed 3-manifold it would be?

If we again consider the first law of thermodynamics and 'Nature abhors a contradiction,' in order to ensure the conservation of gravitational potential energy, we must have it that a three-dimensional spatial manifold on a four-dimensional space is simply connected. If it were not – for example, a 3-torus – a path integral around a hole would give rise to an energy paradox as a particle both gained and lost gravitational potential energy throughout its trajectory, which would be particularly problematic when considering the very early Universe. If the manifold is to be simply connected, we have by the Poincaré conjecture (now proven by Perelman [119–121]) that it must be homeomorphic to a 3-sphere. It seems possible that spatial dimensions are an expanding three-dimensional shell in four-dimensional Euclidean spacetime in which the present is the outer shell.[21] Excesses of matter over patter would cause dents, as if (collapsing three spatial dimensions to two) buttons sewn into an inflating balloon[22] but tethered to a central point by pieces of time-dilating elastic of varying strengths. Spatially expansive ripples could be thought of as poking the balloon from the inside with a cocktail stick, advancing time, before the impulse rippled across the rubber surface: this would make the expansion of space and the onward march of time two geometrical facets of the *same process*. As matter and patter equilibrated, the shells would become more 3-spherical until perfectly so at maximum entropy, when matter would cease to have any gravitational effect. The four-dimensional ball would contain a 'pattersphere' of nodes connected by edges projected from the join semi-lattice. The very early such structure would not be significantly diluted by extensional adiabatic interactions: we might expect to see it projected in the Cosmos as large-scale structure. The holographic principle under this interpretation would accord with the idea that $S = P$, as it would require the information content of the four-dimensional pattersphere to be always encoded in the shape and

---

[21] If this is true, then the parallel postulate of Euclid [49, p. 7] does not hold within the Cosmos over Cosmic distance scales, even in the absence of gravity.

[22] Or perhaps more realistically, non-isotropically expanding foam.

contents of the three-dimensional enclosing spatial manifold as it expanded monotonically.

Under this cosmological model, some thorny problems seem surprisingly easy to deal with. Thinking again back to the very early Universe, the whole structure might be expanding globally, but locally it would be constantly teetering on the brink of gravitational collapse. One might expect this to be an environment full of primordial black holes (but rather more massive than those conjectured by Hawking [67]), but those black holes would contain gradient singularities that could be dealt with by some sort of surgery on a Ricci flow [65, 120] no more frightening than a particle singularity interpreted as the subject of a QED renormalisation, where the former regarded as macroscopic geometry may reduce to the latter microscopically. A link to quantum renormalisation is suggested by Perelman in [121], who also observes some connections to black hole thermodynamics. In our theory, the intense interactions would lead to a constant lessening of their gravitational intensity to the point that they might lose their event horizons surprisingly quickly; a large amount of atypically 'light' matter might then be expected to achieve relativistic speeds at low kinetic energies and have little difficulty escaping into intergalactic space. By the same token, supermassive black holes at galactic centres (which could be some of the same primordial black holes) would tend to lose their gravitational grip, and galaxies would tend to disintegrate with increasing liberature. It would be quite ironic if intense interactions inside black holes turned out to be the main drivers of the spatial *expansion* of the Cosmos. In a universe in which all particles were topologically connected through a join semi-lattice, quantum effects would not seem so very peculiar, all particles being localised only on interaction and continuing to be connected through past time through the pattersphere. Extensive propagation of matter and energy through space would just be the causative mechanism that affected the probability (and hence information content) of diabatic interactions. Elastic interactions consistent at a quantum level with the adiabatic theorem would cause no change in liberature, but redistribute extensive entropy and temperature.

A potential problem in our theory is that the observed quantity of dark energy appears on the surface to exceed that which our theory predicts. This is not necessarily fatal to it. To allow ourselves a little speculation, one possibility is that we were wrong about the Boltzmann constant, and the various factors going spare in our dimensional analysis have some other role, or are related to some yet unanticipated scale factor. A more promising and plausible explanation might be that our view of G turns out to be skewed by our local patter history, which might be at equilibrium locally, but on a Cosmic scale, atypical. Kinetic leakage into gravitons might also have an effect. It is perhaps possible that the underlying G with zero patter might be a 'nicer' number – like unity – related to the curvature implicit in the tightness of the 3-spherical shell and the age of the Universe,[23] and conceivable that our estimation of the amount of dark energy is actually a proxy for the average liberature of the Cosmos. This would carry a potential implication that there is a large amount of mass we did not know about that has already been gravitationally 'neutralised' and we could be well past the halfway point in Cosmic history in *intensional* time. That would be something of a shock.

Where might we find evidence to support a theory such as this? The rate of expansion of the Universe and variations in G would be critical. Deep space observations of gravitational interactions could be expected to help, as may features of radiation that might be expected from dying black holes. Any long-term drift in the mass of ordinary celestial bodies would be difficult to observe as the rate of change would be small on observational timescales, unless effects involving suitable standard candles could be identified; hotter and more intensely interacting celestial bodies would tend to get lighter faster than cooler ones. Terrestrially, direct observation of microgravitational waves from strongly interacting matter might be possible, especially if combined with investigation of any 'leakage' from Planck's law, and evidence of low liberature may be associated with Cosmic rays: the kinetics of interactions of particles of differing liberatures may be anomalous. There is some evidence of black holes that are surprisingly small [145],

---

[23] The apparent ratios and dimensional relationships between κ and the approximate energy, spatial volume and age of the Universe are quite tantalising. However, there are dangers in getting too carried away, as Eddington did, with arbitrary number patterns.

which is tantalisingly suggestive of more rapid evaporation than might be expected by Hawking radiation, but not proof of any particular explanatory theory. However, the most promising avenue for experiments is perhaps in any changes observed in the inertial mass of spacecraft or relatively small celestial bodies subjected to strong interactions. In this case, fly-by anomalies are interesting, and compatible with our theory. If a spacecraft were to pass by Earth inside the magnetosphere it would tend to have a heightened interaction with the Solar wind – especially so nearer the poles, which would result in increased patter, decreased liberature and suprarelativistic mass, and greater than expected increase in observed velocity. Such an effect has been observed and is unexplained by general relativity [5]. Although the speed increase in fly-by anomalies seems relatively high for the effects we are interested in, we note that the Sun is the most potent source of interactions in the Solar System, not just through nuclear fusion but in the constant emission and absorption of photons, and its emissions seem likely to have a correspondingly high patter burden. The surface area to volume ratio of the Earth as compared to a spacecraft would be consistent with a negligible effect on Earth over human timeframes. It is also perhaps *conceivable* that the apparent – although quite weak – correlation in attempts to measure G with length of day variation [6][24] could be connected to a linkage between geomagnetic field variation and Coriolis effects in the Earth's outer core, assuming the terrestrial liberature equilibration rate to be a function of the stable macroscopic material composition of the Earth. Earth, it might be supposed, would be getting lighter (mainly heliocouphically)[25] predominantly from the *out*side *in*, while getting cooler (geothermally) predominantly from the *in*side *out*. However, such an effect would be small and difficult to measure and may be mixed with other causative factors. These potential explanations are intriguing possibilities, although much more investigation is necessary. Perhaps some sufficiently sensitive variant of Cavendish's experiment [29] repeated on spacecraft venturing close to the Sun and in thermal

---

[24] There is some disagreement about how significant this effect is, or what the best fit to the data might be [7, 122].

[25] Neologism (we believe) after ἥλιος and κοῦφος.

equilibrium in deep mines might yield marginally smaller and larger values of G respectively. It is also possible that the close approach of 'Oumuamua to the Sun might have a similar explanation for its anomalous velocity increase, which is unexplained [111]. Finally, we might expect observations of the very early Universe to exhibit double vision, parallax or other distortions as a result of expanding 3-spherical geometry and expansive effects radiating from black holes.

A PECT theory of quantum mechanics could be reconciled with general relativity and thus make quantitative predictions. In Chapter 5, we defined a PECT theory as a category of topologies, and this suggests that general relativity makes a better starting point than quantum mechanics for a PECT theory of quantum gravity, as does the ease with which any lattice can be interpreted as a topological space. A general approach to this might be guided by extrapolating from Wheeler's soundbite: 'Space tells matter how to move. Matter tells space how to curve.' [157, p. 235] If the interpretation of dark energy we have described is correct, then we would have the following: 'Extensive space tells matter how to propagate. Matter tells SND how probable interactions are. Actual interactions tell matter how to change. Matter and patter tell extensional spacetime how to curve at interaction-time.' This gives some idea of how one could start to hack the Einstein field equations [47] to integrate such a theory of quantum gravity without a cosmological constant. The relativity and quantum field elements of these equations would be deterministic, massively parallel, and perfectly and exquisitely *analogue* – not digital – to an unbounded precision; the information element would be non-deterministic. The ultimate origin of information that drives non-deterministic choice and intensional time is outside the scope of the theory and of physics. We note, however, that nothing requires this cosmontic information to be determined before (in intensional time) the moment at which it unfolds extensionally, and therefore does not require intensional determinism or endorse Leibniz, Spinoza or indeed Dr Pangloss. We shall address the question of the compressibility of cosmontic information and what we might be able to infer about it inductively in Chapter 7.

The result of such a join-semilattice theory of dark energy as we have described would result in an extensional world in which information about history is constantly *convolved* with new information. If the theory is correct, then enlightenment seems to be unveiled as things become lighter and more like light. It would seem according to these rules to be fruitless for us to try to pluck knowledge of the state of all Nature from the Bayesian join-semilattice of which all life is a part.

The development of a topological theory of the kind we have described forms a subject for further work. Whilst it does not currently make complete quantitative predictions, even the qualitative theory admits an avenue for experimental investigation.

While we would be surprised if this theory is right in every respect, we shall perhaps be more surprised – given its simplicity and symmetry – if it turns out to be entirely wrong. Thinking on Watts's metrical version of Psalm 90, time might bear all its sons away *extensively* in a local frame of our choosing; a suprarelativistic theory of hierarchical particle dynamics would suggest that even *physically extensionally*, all of substance that has ever been in the Cosmos, still is, both *intensively* and *extensively*. *Intensionally*, we shall – with perhaps a twinkle in the eye – say nothing for now. There is a little *coup de théâtre* to come. But first, we need to do some more work.

6.2   Prescriptive theories

Any type of engineering can be viewed as in essence the business of reifying a prescriptive theory: an engineered system has some 'input' degrees of freedom and given what transpires within those degrees of freedom, it should behave in some particular way. This is as true about the steering wheel of a car as the buttons on a pocket calculator or a web application: discrete or continuous, static or dynamic, the principle is the same. The PECT theories we are interested in in this case are of *machines*. If we cannot say what will happen within some equivalence class of histories of an engineered system, then we do not have a complete specification. An AI system that purports to be able to control a system capable of causing harm is only completely specified if it either specifies what happens when faced with every identifiable ethical dilemma or explicitly states through an equivalence class that the specifier does not care what happens for some range of intractable scenarios. It seems reasonable to suppose that requiring a specification to meet a completeness criterion would force such hidden and often inconvenient issues to the forefront and ought to prevent misconceived systems from leaving the drawing board. *Implementing* an engineered system is the process of arranging a weak bisimulation of the prescriptive theory by supplying an appropriate configuration of matter to a complete physical theory at a suitable level of abstraction.

Normally in engineering specification, a 'system' is the primary entity and a specification prescribes its behaviour. In the $I'V'T'E'$ ontology, it is the S_L space that is the primary entity, and a specification is a set of constraints on a space *witnessed* by some system that it *admits*, which may *trace* or *describe* the specification as it evolves through time if that system remains *correct*.

In Section 5.3, we noted that the subset of labels S_L over which a specification may range is a dynamic entity and function of epistemic time and its *history* until that time. Specification over an S_L subspace is not therefore analagous to defining a Hamiltonian over S_L.

To program a discrete $I'V'T'E'$ ontology, we take the position that anthropontic 'data' (as a matter of definition) exist if and only if a natural person (or persons) has, directly or indirectly, intentionally instantiated them as a set of coordinates in a coordinate system of *identifier*, (discrete) *value* and *time*. To state general engineering specifications,[26] we can accept equivalence classes of real-valued extensional configurations of matter within tolerances, as discussed in Section 5.2. In this case, the values do not refer to non-spatial abstract assignments to labels by fiat, or the transitive consequences of fiat, but – although again by fiat – to a prescribed configuration of identifiable physical items with world lines. Engineering and computing specifications should be preservative over their *histories* even if they are not preservative from one state to another. Data may *appear* to be erased but only because we cannot control the degrees of freedom through which

---

[26] Which can be analogue computations if we input analogue data and are prepared to accept our answers as partially non-deterministic equivalence classes.

new (if the process is diabatic) and old information is convolved with the wider state of the Cosmos. The flip side of this is familiar to cryptanalysts as the concern about *side channels*. Prescriptive specifications are grounded in physical reality in the extensional $I'V'T'E'$ of a physical PECT or a homeomorphic abstraction layer built on top of it. In a computational specification, the only dimensions that are tied to physics are time axes: intensionally, E is aligned with it; extensionally in a continuous specification, T' is tied to it *contingently* and E' is tied to it *actually* subject to 'noise' in E, either physical 'noise' information in the *physical E'* if the level of abstraction is $\mathbb{P}$, or noise that is 'summarised' at an appropriate level of abstraction. In an engineering specification, physical spatial dimensions are embedded in V'. The subset of labels L, $L \subseteq I$ and of labels L', $L' \subseteq I'$ of a particular specification, along with other constraints, can be specified using predicates in a suitable theory, as will be discussed in Section 6.3. The meaningfulness of these predicates depends on an external ontology, implementability in a theory at a lower level of abstraction, and correspondence of entities between levels of abstraction and with identifiable entities in the relevant ontology.

The requirement for a specific ontology, built on a general one we are about to elaborate from Section 5.2, is born of requiring an anthropontic *pragmatics* for a theory. In a serious sense, as we introduced in Section 2.1, specification or program correctness is meaningless without one: to rephrase Babbage, if one does not know what the machine is supposed to compute [11, p. 67], how can one know whether what comes out is right or wrong, other than perhaps by interrogating Arthur Dent? [4, p. 149]

6.2.1 Ontology of time and objects

Before we go any further, we need to constrain our model of extensive time for prescriptive theories. Curved spacetime is not very helpful for most engineering purposes, and even when necessary in space engineering, it does not need to be part of the fabric of the axiomatisation. For most engineering, a Euclidean model of spacetime is perfectly adequate, and does not introduce practical difficulties when we want to bisimulate (weakly) the specification in real, physical spacetime through some causation-invariant homeomorphism. Some concept of simultaneity is by definition essential for anything that is supposed to be synchronous. Usually, this is arranged with reference to some arbitrary clock, but a coordinate timescale is more powerful. If we are to *compose* systems, we need a common timescale, and the obvious relativistically consistent common denominator for terrestrial engineering is TT or Terrestrial Time [76, Rec. IV] (defined as Geocentric Coordinate Time [76, Rec. III] with a fixed time dilation and offset), which is well approximated by proper time anywhere on Earth's geoid.[27] This timescale is effectively witnessed by the clock ensemble of *Temps Atomique International*.[28] In order to deal with systems that cannot be defined relative to a global time standard, or where it is inconvenient, such as among a constellation of satellites, we need a formal notion of time composability.

A physical theory $\mathbb{P}$ may have a set of primitive types of point in $I'V'T'E'$ space that can be identified by partitioning I' and embedding potential degrees of freedom in V'. Equivalently, static properties may be encoded in V' along with the dynamic ones. A prescriptive theory is no different, except that the identification of what points represent is some human judgement associating the anthropontic with the hypokeimenontic, rather than trying to learn how Nature associates the cosmontic with the hyperkeimenontic. In order to specify a machine, some particular $I'V'T'E'$ space is needed as a canvas for the specification.[29] This is a synthetic, extensive canvas, and there can be as many of them as there are specifications. The nature of such an $I'V'T'E'$ space has two aspects. The first can be formalised in a system of logic (which is identified in natural language), and encompasses the structural properties of identifiers, values and time in the $I'V'T'E'$ space. We consider systems of logic in Section 6.3. The second cannot be formalised, and concerns the human or organisational root of authority for determining

---

[27] A similar idea was first presented in [56].

[28] McCarthy and Seidelmann give a comprehensive history of the somewhat intricate history of TAI [100].

[29] We omit discussion of the IVE parameter as it follows the observation semantics already discussed.

what real-world objects or concepts the identifiers refer to, and the inertial frame of the timebase. The most formal this second kind of property of an $I'V'T'E'$ space can be made is that it be contained in a natural language document with a unique identifier or by some convention established in such a document. This includes a predicate specification over a non-finite set of labels in I' that encode the rules of some particular ontology external to the $I'V'T'E'$ ontology: such an ontology is itself hypokeimenontic and can be encoded anthropontically.[30] Such an ontology can also establish predicates over the V' that can be associated with labels in I'. In the case of a physical piece of engineering this might include some solid geometry, material composition, and rotational and translational kinematic parameters from a suitable palette. We can simulate a specification over an $I'V'T'E'$ space using just the structural properties, but uniqueness and composability as hypokeimenontic *engineering* properties come from the second type of property; the natural way of specifying them is in international standards. International standards usually specify types of entity, but there is no reason why they cannot reference particular entities of types through iso/itu object identifiers (oids) [144].[31] We define two $I'V'T'E'$ spaces S_1 and S_2 to be *weakly composable spaces* if $I'_{S_1} \cap I'_{S_2} = \varnothing$ and *strongly composable spaces* if additionally T'_{S_1} and T'_{S_2} refer to homeomorphic timebases in the same or different inertial frame (a structurally similar idea to radar time [22, p. 26]),[32] with the same thing able to be said about the corresponding E's. Note that the T's and E's in this case have to be identified in natural language or by some convention established in natural language. Note also that the structural properties of V', T' and E' in an $I'V'T'E'$ space do not affect the composability of the $I'V'T'E'$ *spaces*, but the structural aspect of the V's, T's

---

[30] As long as there is a base case document somewhere whose retrieval is not subject to the functioning of a machine. One could view a formal relation of the hypokeimenontic to the anthropontic as a kind of higher-order hypothetical Aristotelian hylomorphism, except that we allow the 'hylomorphism' itself to be hypokeimenontic.

[31] This kind of idea was developed in [56].

[32] The idea comes from Milne, before the acronym 'radar' was coined – his invocation of Morse code is germane to our use [103, p. 27].

and E's must either be the same or have defined relationships in order to be able to specify constraints in general on $S_1 \cup S_2$. Such spaces can be considered within a common logical denominator in the manner discussed in Chapter 5, within the framework of Section 6.3 and Chapter 7.

6.2.2 Continuous machines

The most general kind of engineering specification is a continuous, real-numbered one. This includes everything from parts of a bridge to the layout of physical features on a silicon chip. Time and space are real-numbered in such a specification, and the next state is generally determined by integrating an action with respect to time based on the current state at each time. This is not quite the whole story, because sometimes it is necessary to specify something over a history or in a frequency domain: for example, a bridge resonating with the march of an army's boots, a battery whose charge and discharge cycles must be constrained to prevent it from overheating, or a mechanism that must be operated in some particular order or at a particular speed to avoid mechanical damage.

In general, the behaviour of such a system is obtained by integrating monolithically the semantics of its actions with respect to time, because components that act on one another have inertia or impedance, and generate reactions or reactance. Those semantics will be some intermediate abstraction on $\mathbb{P}$, such as analytical mechanics for mechanical systems. These semantics are always an approximation, as if a piece of engineering is to be *reliable*,[33] it must somehow maintain a predictable configuration modulo an equivalence class of immaterial difference over some interval of time, and do this *in spite of* the diabatic entry of natural cosmontic information from quantum processes. We shall return to this issue in Section 6.3, but first we consider discrete state machines and their semantics, which is the traditional domain of computer science.

---

[33] A definition of engineering reliability usually reduces to some idea of correctness, but as we established in Chapter 2, nobody really knows what that means – which has led to the formal idea of complete specification we have given in the first five chapters of this book.

6.2.3 Discrete machines

A discrete state machine is one that must inevitably be implemented over a continuous one: it seems to be a law of Nature that we do not have access to Nature's 'pure' information, even supposing it is intrinsically discrete. It is implicit that discrete state machines deal with discrete data, but what about time? Lamport's model of events in [88] seems close to Nature, and some kind of discrete advance of time is the best we can do when manipulating discrete data. Even if Nature has discontinuous episodes, we have at best an approximate idea when they will happen in continuous time, which according to Section 6.1, might be suspended for the duration. There is merit in adopting a dense time formalism of rational numbers for full flexibility, of which the Harmonic Box Coordination Language (HBCL) of [56] is one. A paper giving a revised language with a PECT characterisation of its semantics is planned. It was the requirement for a semantic formalism that allowed specification over a complete *history* that led to the present work, since that is the most natural way of expressing a first-in-first-out (FIFO) pipeline, which is the basic unit of latency in HBCL. A fully discrete timeline is another possible choice, and in any case a particular specification in a dense time formalism must reduce to some minimum clock rate. We now therefore consider what the specification over S_L and the space itself should look like for a discrete machine.

Programming languages have semantics: static semantics determine the set of valid programs given some abstract syntax, while dynamic semantics dictate how the abstract state space of the program evolves as it is executed. All programs are banally complete, but not necessarily in a way the programmer intended. They can though, when demonstrated to obey the static semantics, prove that an equivalent predicate phrasing of the specification is executable. Neither sort of semantics can determine how the abstract state space of the program relates to the real world when it is compiled and executed or interpreted by some abstract machine. Giving a pragmatics in this paradigm would involve leaking abstractions from particular machines or platforms, at which point one is no longer really dealing with one language:

something like this can happen rather vaguely in languages with pragma directives. The 'reality' of data is always contingent on some ultimately binary representation, manipulated in the case of a conventional von Neumann architecture [113] according to the instruction set of some processing unit. An alternative solution is to throw away the hardware and software distinction completely and concentrate on defining what happens to data according to some abstract reality, without getting too caught up in the nature of computability. It does not matter much what the philosophical status of that reality is, as long as it is sufficiently clear that it can be consistently conceptualised by a human mind.[34] By making a programming language *parametric* on such an ontology, we can clearly separate pragmatics from semantics, eschew the implementation-dependent but rather artificial distinction between hardware and software, and bake into a programming or specification language the requirement that it be clearly instantiated.

A programming language which, in a limited way, allows programming with respect to history, is a good way to express something akin to functional reactive programming (FRP)[35] [48] paradigms but with deterministic time delay. However, in order to be implementable, it must be transformed into a form in which history that affects the future is memorised and preserved in a state, since we do not have access to patter (if it exists) and our implementations are not isolated from the rest of the Universe. In the discrete world, reality has the peculiar feature that there is no quasi-instantaneous feedback: action is determined by entities with effectively infinite impedance over entities with zero impedance.[36] Between information entering the system through an IVE parameter, we can determine the extensional evolution of the $I'V'T'$ subspace for a fixed e' in E'. First, we can

---

[34] The shift from hardware to software from ENIAC to the Manchester Baby (for a history see Lavington [92, pp. 273–298]) might have been useful but the difference looks more slight than the philosophically quite rigid hardware and software distinction we have today.

[35] Originally conceived by Elliott and Hudak as 'Functional Reactive Animation', but 'FRP' has stuck.

[36] This is not quite the case with hardware description languages (HDLS) such as VHDL, but that is more due to HDLS being slightly leaky abstractions.

divide S_L into S_{L_∞} and S_{L_0} for the purposes of defining a next state relation; the two spaces may overlap. The points in S_{L_∞} (indexed by L_∞) represent a virtual world line with infinite impedance: they are parameters constrained from outside S_L or by a previous state transition. The points in S_{L_0} (indexed by L_0) are created according to the specification. S_{L_∞} may be empty, in which case the specification is closed, and the points it constrains are a function only of time. A specification on S_{L_∞} and S_{L_0} as indexed by the Ls at different times in E can be encoded in a binary relation on the respective supersets of these S_{L_∞} and S_{L_0}. However, allowing arbitrary specifications of this kind includes specifications that are non-deterministic and violate causality, so some subsetting of this relation is clearly required to determine the properties of an abstract specification in an $I'V'T'E'$ space.

Given that we want a *programming* language to be deterministic, we need some extra equipment to state the minimum properties on an $S_{L_\infty} \times S_{L_0}$ relation (we elide here any variation in L with E for simplicity of exposition). Since this is cumbersome in the general case, a given *specification language* should guarantee that it holds for the set of valid programs that it admits. It can do this by requiring the totality, soundness and surjection conditions of formulae (2.1) and (2.2) and a carrier for temporal specifications that axiomatises causality and monotonicity (formulae (4.5) and (4.6)). For a computation not explicitly handling faults, all that is required is to plug S_{L_∞} and S_{L_0} into D and C respectively for formulae (2.1) and (2.2) and $\mathcal{D}''$ and $\mathcal{C}''$ respectively for formulae (4.5) and (4.6), fitting the $I'V'T'$ of the S_L space to W as $(I' \rightharpoonup V') \times T'$, flattening label into value as a partial function in a Cartesian product with time. Note that a label can be both in the L_∞ and L_0 sets but according to formula (4.5) cannot be bound by a specification F' in both S_{L_∞} and S_{L_0} at the same instant in T'. For static configurations, the projection of $I' \rightharpoonup V'$ on V is restricted to be constant for all T' (and T). Time is considered to be distributive across I', axiomatising simultaneity. This is a reasonable semantics in an ontology with zero spatial dimensions and relativistically consistent if, when projected into reality, all observations are timed axiomatically according to an observer at a single point in spacetime, which is the case for a coordinate system axiomatised in Terrestrial Time.

6.3 Logical theories

The axiomatisation of logical theories is riven with circularity and infinite regress. Harrison sums it up in *'quis custodiet ipsos custodes'* [66, p. 1][37] or 'who shall guard the guards themselves?' Metalogic can be viewed as just an abstruse kind of logic, and in what does one axiomatise the metalogic anyway? Larger and larger inifinities are needed, as Turing found in [153]. Somehow, without falling foul of Gödel, the truth seems to reside in a fixpoint of expanding infinities, but how does this relate to reality?

The problem is easier to tackle in the context of a formal logic itself axiomatised in a formal logic, in a way similar to that described by Harrison [66], Davis [36] and Myreen [110] (surveyed in [56, pp. 52–54]). The trick is usually to keep the extra axiom needed to show consistency small and credible, which turns out to be to assume that some larger and more exotic infinity 'exists' than can be handled *inside* the theory. However, this is philosophically uncomfortable because the extra-logical argument to justify the axiom needed seems anthropocentric and often leads to some very woolly and mystical reasoning about the nature of human minds and their willingness to accept the reality of tangible infinities. This kind of infinitistic axiomatisation of reality troubled many nineteenth century mathematicians who were uncomfortable with Cantor's infinitistic ideas, but since Russell, Hilbert and Gödel these concepts have not been considered controversial in the mathematical mainstream.

We believe it is more useful to view the consistency of logic as a conditioned emanation of the apparent uniformity of physical laws across time and space. This is hardly a new idea, probably going back in some form at least to Aristotle *via* Aquinas, but we can understand it with a Darwinian spin [35]. The Curry-Howard isomorphism [140], which equates computations with proofs, provides the insight with which we can view a logic as a PECT theory. Theorems in a logic are *tautologies*: unless they have some semantic model, they *reveal* no new information that was not inherent in the

---

[37] This well-known and apposite phrase originates (regrettably) from the abominable Sixth Satire of Juvenal.

logic, and in any event *contain* none. A proof has an information content, but confirmation that the proof is indeed a proof contains no more information than the proof itself and the rules of inference of the logic. The consistency of logics in the intuitionistic type theory paradigm [97] is axiomatised as *strong normalisation*. In this paradigm, function types state propositions and proofs are functions that inhabit these types. The consistency of the logic is based on the rewriting of proof terms into a normal form according to the rules of the logic, such that any well-typed term will always reach the same terminal state by applying the rewriting rules. Intuitionistic formalisms are the first-class citizens of such an approach to defining logics: other kinds of logics start with *a priori* assumptions, which cannot be grounded in evidence until they are proven to make predictions consistent with the way the Universe actually is.

The formulation of consistency as strong normalisation is not a magic answer to the question of demonstrating the consistency of a logic in itself, but the fact that we can check a logic mechanically suggests that we might be able to replace logical axioms with physical ones if we consider the logic in terms of its scalable physical checker. We now examine a way in which a logic $\mathcal{L}$ in a family of logics $\mathbb{L}$ can be induced by $\mathbb{P}$ without recourse to infinities if we are content to restrict our idea of consistency to $\mathbb{P}$-consistency. The class of machines we shall be interested in for checking consistency cannot be Turing-complete, because we want them to terminate (for finite inputs). We can axiomatise this for some given subset of labels $L' \subseteq I'$, as there existing some extensional $t_{\max} \in T'$ for which for all $t, t \geqslant t_{\max}$, V' is constant.

Any deterministic machine can potentially be viewed as a model of something other than itself. For a logic, we are particularly interested in machines that encode a proposition, a purported proof of it, and have an end state that can be labelled 'true' or 'false.' For the logic to be any use, it must also be syntactically consistent – that is, it should not be possible to prove both proposition P and $\neg P$. We obtain this in an intuitionistic logic by requiring strong normalisation – a second mode of a putative logic machine *reduces* proof terms to a normal form if and only if the first mode (the type checker) evaluates to 'true.' Perhaps this would be more of a catalytical engine than an analytical one. Any logic that provides the language for propositions and proofs that can be so encoded has a satisfiable model in the machine, because we *started* with the machine. The requirements of two modes of the machine prescribe that type checking be decidable for any logic that satisfies them. We also have the problem, common to the 'institutions' we met in Chapter 5, that our propositions and deductive apparatus could be so banal as to make it impossible to embed additional meaning within it. If we require of this class of deterministic machine that it be capable of encoding itself according to the axiomatic physics of its deterministic theory and the encoding of its propositions and proofs (a double encoding), this provides a minimum of expressive power for a logic. We can also require that for all such representations, we should be able to prove a morphism between the logic embedding and physical termination. If we can demonstrate physical termination for the reduction mode of the machine *provided* the type checking mode yields 'true,' we can show consistency *up to* the consistency of the physical machine. If we admit the set of all such logic-machine pairs definable in – or equivalently *induced by* – $\mathbb{P}$, we have a maximum expressive power of logics accessible to us. These requirements make some minimum immediate demands of a logic: for example, it must admit coinduction in order that a termination proof – of a point after which a cofixpoint representing the evolution of a physical system reaches a fixpoint – can be given.

There is a wrinkle in this idea, which is that, from Section 6.1, we have that $\mathbb{P}$ has both deterministic and probabalistic elements. This suggests that we need somehow to restrict $\mathbb{P}$ to a deterministic form to have 'certainty' in our mechanistic logical deductions, something that the quantum aspects of $\mathbb{P}$ would tell us is impossible in a less than perfectly adiabatic system. In order to address this, we turn to two classical results in computer science on reliability, since through the Curry-Howard correspondence confidence in knowledge is isomorphic with reliability in computation – this is very obvious in the case of quantum computing but less so in ordinary computing where we are accustomed to high levels of hardware reliability. The first piece of good news is von Neumann's result on probabilistic logics [112], that an arbitrarily reliable machine can be made from arbitrarily unreliable components. We might not be able to have total certainty, but there is no upper limit on the level of certainty we can have until we run out of resources, inclination or sense. We also have a similar result from Shannon's noisy channel coding theorem [136, p. 411], which tells us that information can be coded to make the probability of error in information storage or transmission as arbitrarily small as sates our appetite, for some given level of noise. We can therefore build a contingently deterministic class of machine $\mathbb{M}_{x \cdot \sigma}$ embedded in $\mathbb{P}$, where x quantifies the level of confidence required. We perhaps should not be surprised that our level of certainty must be limited in this way, since the science that allows us to build such a machine with any confidence is necessarily inductive and corrigible in the first place. Once we have confidence in the principle of $\mathbb{M}$, we do not necessarily have to build a big enough catalytical engine[38] that proceeds *actually* to carry out the second mode for large interesting theories: if we are prepared to accept an axiom of physical scalability,[39] it only needs to be built at a scale sufficient to process the size of the proof of interest without indigestion.

Given that we have said a PECT theory is a topology, how can we make a logic into one? The most elegant answer to this question can be found in homotopy type theory [155], which provides a bridge between intuitionistic type theory and topology. This sets up the possibility of a univalent[40] mutual embedding of physical and logical theories as we shall discuss in Chapter 7.

---

[38] Or find enough spare information encoding capacity in the Cosmos.

[39] For *coinduction* without limit, the scaling would have to be unbounded. However, from inside the Cosmos, intensional time appears to be bounded by heat death and before that by the fact that there is in any case insufficient material within the Cosmos to simulate itself – for obvious Gödelesque reasons. In order to frame termination (and hyperkeimenontic heat death of $\mathbb{P}$ itself) unrestricted coinduction seems necessary, which aligns with an interpretation of a 3-spherical universe in four-dimensional *Euclidean* spacetime. Such an imaginative step seems more conservative than a belief in the completed Platonic existence of ε numbers.

[40] This term was originated by Voevodsky in the context of his univalent foundations of mathematics programme [9].

Chapter 7

Transfinite induction and the scientific method

So far, we have introduced the idea of a PECT theory, and discussed the properties of a PECT physical (a subset of *de*scriptive theories), *pre*scriptive and logical theories. In Chapter 3, we gave a definition of computability that involved physics and logics; in Section 6.3 we saw how a logical theory depends on a prescriptive definition of a verification machine, which is a special case of the kinds of engineering specification we discussed in Section 6.2. Throughout, we have observed the circular relationships of these ideas, and suggested that experimental induction through a physical theory is the way to break the circularity. In this chapter, we shall put all this together, suggest how it can be assembled in a univalent topology, and briefly discuss the abstract features of such a topology. We might seek some inspiration from a dictionary: its meaning seems to inhabit *structure*, given that every word in a single language dictionary is defined only in terms of other words we should be able to find in the same dictionary. We shall conclude our adventure with some final observations about thermodynamics, agency, and actions that might – among other things – increase our confidence in a theory.

If a PECT physical theory is isomorphic with a computable function, then according to equation (3.1), the definition of an extensional version of it for a $\mathcal{P}^\sim$ (fitting $\mathbb{P}$ as perfectly as $\mathcal{P}$ *but for* scale or boundedness) should satisfy

$$\forall d \in D.\ \forall c \in C.\ \mathcal{P}^\sim (d,c) \Rightarrow$$
$$P\left(Y(c) \mid X(d) \cap Z\left(\mathscr{E}_{i\mathcal{L}_{\mathbb{P}^\sim}}(\mathcal{P}^\sim)\right) \cap \mathcal{L}_{\mathbb{P}^\sim} \cap \mathbb{P}\right) = 1 \qquad (7.1)$$

where $\mathcal{P}^\sim$ is a scalable set-theoretic phrasing of the extensional function of the Cosmos, $\mathcal{L}_{\mathbb{P}^\sim}$ is a $\mathbb{P}^\sim$-consistent logic, d is the intensional history of the Cosmos and c its extensional history.[1] $\mathcal{P}^{-1}$ is also the compression function for the extensional Cosmos to its information content given the uniformity of physics. We note that, regardless of whether $\mathbb{P}$ is intensionally discrete, $\mathbb{P}^{-1}$ is *not* a total function unless it is limited to real-numbered histories that are possible given possible intensional 'inputs'; if $\mathbb{P}$ is fundamentally intensionally discrete – which seems to be hinted at by quantum mechanics – the only generators of such histories are Diophantine solutions to such possible extensional worlds.[2] The question now arises: if we dare to map the hypokeimenontic modelisation to the hyperkeimenontic of the physically real, what do we have to do in order to have high confidence, rationally, that a $\mathcal{P}^\sim$ satisfies equation (7.1) and (but for scale if $\mathcal{P}^\sim \neq \mathcal{P}$) is a member of $\mathbb{P}$?

The only rational tools that we have are the *inductive* approach of the scientific method, which axiomatises our conditioned quotidian belief in the uniformity of Nature. The scientific endeavour that establishes, but does not *formalise* a $\mathcal{P}$ is the sum of physical experiment and the consequent body of knowledge built up by physical science. The consistency of a formalised theory with the known results in the equivalent physics couched in everyday mathematics would help to establish its correctness. However, particular subsets of it could be made more convincing by the self-axiomatisation of a proof checking machine in the way we began to discuss in Section 6.3. We suggest that an axiomatised theory of physics, a logic and a verification machine whose structure meets the constraints we have given would make a convincing case that all of them were consistent, as the

---

[1] A very early precursor to this fixpoint appears in [56, p. 46].

[2] It would seem that this must be the case if a perfectly analogue extensive universe is to contain a bounded amount of monotonically increasing information.

chances of compensating mistakes masking an error seem small. The confidence in such an argument can be increased indefinitely by swapping in new $\mathcal{P}^\sim$, $\mathcal{L}_{\mathbb{P}^\sim}$, $\mathcal{L}_{\mathbb{L}_{\mathbb{P}^\sim}}$, the real proof checker $\mathbb{M}_{\mathbb{P},\mathbb{L}_{\mathbb{P}^\sim}}$ (and axiomatisation of it $\mathcal{M}_{\mathcal{P}^\sim,\mathbb{L}_{\mathbb{P}^\sim}}$) and showing morphisms under equivalence relations. However, proof in a $\mathbb{P}$-consistent logic is always contingent and ultimately empirical. The embedding scheme of these theories should have the shape shown in Figure 7.1, modulo a conservative scaling or bounding axiom that $\mathbb{P}^\sim \Leftrightarrow \mathbb{P}$.

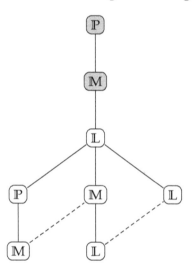

Figure 7.1: Embedding of theories for an induction argument

The solid lines show embeddings of one theory in another; dotted lines show morphisms to be shown in the root '$\mathbb{L}$.' The grey '$\mathbb{P}$' and '$\mathbb{M}$' represent the actual extensional instance of $\mathbb{P}$ that we inhabit and the actual physical proof checker that is checking the proofs we have in $\mathbb{L}$. We neglect subscripts and superscripts in the diagram for clarity. The locus of induction is the belief that the physically hyperkeimenontic reality we are *in* is accurately described by the scaled hypokeimenontic models we have made of it. We note that $\mathcal{L}_{\mathbb{P}^\sim}$ and $\mathcal{P}^\sim$ may be narrower than the maximally powerful $\mathbb{L}_{\mathbb{P}}$ and the complete $\mathbb{P}$ if they have features that are not used in $\mathbb{L}_{\mathbb{P}^\sim}$, accessible in $\mathcal{P}^\sim$ or present in $\mathbb{M}_{\mathcal{P}^\sim,\mathbb{L}_{\mathbb{P}^\sim}}$. $\mathbb{M}_{\mathbb{P},\mathbb{L}_{\mathbb{P}^\sim}}$ differs from $\mathbb{M}_{\mathcal{P}^\sim,\mathbb{L}_{\mathbb{P}^\sim}}$ in that the former is the actual machine and the latter as it is axiomatised, usually within some open fragment $\mathcal{P}^{\sim}$ that lacks knowledge of all $\mathbb{P}$. Once we have one such structure, we can mix and match them, axiomatising more than one subject axiomatisation of some $\mathbb{L}_{\bar{\mathbb{P}}}$ say, within the same comparator theory.

The result of this structure is an *empirical* specification for the proof of a $\mathbb{P}$-consistent logic. We do not fall foul of the incompleteness theorems because we are not trying to prove that $\mathcal{L}$ is $\mathcal{L}$-consistent or even $\mathbb{L}$-consistent, merely $\mathbb{P}$-consistent, where the truth of the axioms of $\mathbb{P}$ are physical laws established by the scientific method. Once we have this inductive base case, we can follow a nested relationship down as many rabbit holes as we please, but it is a *potential* infinity of embeddings, not necessarily an actual one – it ceases to be meaningful when we run out of resources *inside* our extensional $\mathbb{P}$ such that we can take it no further. The key requirement to find credibility in a structure such as this is an appreciation that physics as encoded scales uniformly to physics as it actually is, the machine as encoded corresponds to the machine actually built and used to conduct a 'true'/'false' experiment, and that the logic as verified by the machine corresponds to the logic that we have axiomatised in itself. By scaling an inductive base case, we can avoid the need for fixpoint infinities like ε_0, which is usually required for Gentzen-style proofs of the consistency of arithmetic, let alone anything more exotic than that. Hilbert's second problem [70, pp. 447–449] on the axiomatisation of arithmetic (but only up to $\mathbb{P}$-consistency), the sixth problem on the axiomatisation of physics, and the Church-Turing thesis, can be linked in the fixpoint[3] of PECT theories we have described. The inductive basis for the physical case breaks the circularity and opens a way to find as convincing an account of all three issues as is possible with our rational perception as limited *inside* physics. We come to believe that a hypothetical, hypokeimenontic theory of physics is hyp*er*keimenontic insofar as it is consistent with and *continues* to be consistent with the results of experimental enquiry.

We now seem to have a 'univalent' topological structure, to borrow the

---

[3] For practical purposes inside the finite Cosmos, it is a fixpoint: whether there is a cofix-point in part of an unbounded deeper reality is a metaphysical question. However, we will go on to consider whether science can say anything about the weaker question of whether $\mathbb{P}$ can *possibly* be an *actual* base case or whether such a case (if it exists) is necessarily ethereal.

use of that term from Voevodsky [9]. We have defined a PECT to be a top-ology. Our family of logical theories can be viewed as a topological one through homotopy type theory [155], and our physical theory is also a topo-logical one. A proof checking machine can be defined in the logic and is again a topology. We believe it is a worthwhile agenda to extend the idea of a univalent foundation of mathematics as conceived by Voevodsky with a univalent foundation of physics as a mutually inductive fixpoint – a kind of mutually assured construction. It would produce as a by-product a sub-strate for complete *engineering* specification and serve as a checker for im-plementations derived semi-automatically by repurposing (and preferably renaming) artificial intelligence approaches.

The experimentally inductive nature of $\mathbb{P}$ leads to a further question about scientific enquiry that is not concerned with basic physical phenom-ena. There are two possible categories of such enquiry. The first is the in-ductive investigation of systems that are currently too complex to describe accurately by physical modelling, where in theory a *deductive* solution from more basic knowledge would be possible. The second category of enquiry is that which seeks whether it is necessary to explain experimental results that cosmontic information have a denser structure than appears in a phys-ical theory, not unlike the sorts of dimensions we are encouraged to think about by Abbott in *Flatland* [1]: in other words, is the self-information of cosmontic information smaller than its apparent self-information in a phys-ical theory and perhaps equal to the information entropy of some type of ethereally hyperkeimenontic structure? It seems possible that some phe-nomena of mind, for example, may require self-correlation in the intensional cosmontic information that gives rise to its extensional form in order to ex-plain its outward phenomenology. If free will is real, if we are a 'strange loop' (to borrow a phrase of Hofstadter [72]) *mediated* by the physiology and innate and learned predispositions of belief of our minds, and their extensional surroundings, then social science would be expected to be nec-essarily inductive in order to reveal the patterns in cosmontic information that drive its extensional manifestation. If, however, all the phenomena we observe in the Cosmos can be generated by substituting for cosmontic information a pseudorandom anthropontic number generator, then the 'heterophenomenological' approach of Dennett [38] might be able to characterise the behaviour of systems as a whole, although would not be able to prove that any action of the *individual* was not the result of a hyperphysical aspect of consciousness. If, though, social, creative and aesthetic structures or a moral bias allow further compression of extensional physical information than afforded by $\mathbb{P}^{-1}$, this would virtually prove the existence of a hyperphysical aspect of our reality in some sort of intensional feedback loop from physical extensionality, breaking the assumption of $\mathcal{D}' \cap \mathcal{C}' = \emptyset$ in Chapter 4 more concretely than in Chapter 5. This perhaps offers some way out of the paradoxes and circularities of J. S. Mill [102, pp. 479–489]. There seems to be a hint of this in that coordinated human activity through some collective intentionality [135] can produce *voluntarily* a physical spatial or temporal periodicity that apparently reduces the *global* intensional information content of the naïvely *local* causative influences: making music together, both literally and figuratively. A conductor who puts down a beat does so with a faith in the *good will* of musicians in the moment; an engineer who presses 'play' does so with a faith in the quality of some engineering in *past* time. Musicians on a platform are invited to be co-creators in something original and beautiful, not browbeaten into servility like machines.

What we find convincing *rationally* may be conditioned by the uniform aspects of Nature. However, the extensional reality of non-determinism induces a limit on the *certain* knowledge we can have of the nature of consciousness and autonomous agency, and speaks to the folly of attempting to answer all questions through powers of reason: a prevailing Procrustean proclivity to wrest everything into the digital dust of a *reductive* reason and intellectual *cul-de-sac* of sterile phenomenological nullity. We are free to infer knowledge through other aspects of experience pertaining to reality through induction, introspection, insight, inspiration, observation or teaching of a kind that is outside the realms of *scientific* enquiry. It is difficult to know how else to go about finding answers to such questions.

If we recall that an extensional causal theory contains no information not implicit in the theory and its parameter, we might imagine that there were no reality outside the physical: then all human knowledge would be illusory – a mere phenotypical emanation of that intensional information content whose apparent intricacy would be no more remarkable than the apparent extensional complexity of the Mandelbrot set [44]. However, to proceed directly from a distaste at this possibility to a hyperphysical dimension to reality is to succumb to a teleological argument.

With this in mind, other than some proof that an observed small-scale phenomenon cannot be explained without self-correlation in the parameter to a physical PECT, is there any other way to illuminate the question of possible correlation *inter se* in intensional information? The second law of thermodynamics appears to offer an answer: if the agency of mankind were necessarily to emerge from a PECT reality that could equally have been driven by incompressible entropy[4] or the output of a pseudorandom number generator correlated only by the algorithm that defines it, we must conclude that the ways in which, on a very large scale, humanity disturbs equilibria violate the second law, albeit at the expense of creating more entropy overall. The second law seems to be one of the most secure laws of Nature that we have inferred, even if the tendency of things to equilibrate – regression to the mean or reduction to mediocrity – is often mixed up with entropy increase, with instantaneous indistinguishability recruited as the phenomenological resolution to Gibbs's 'paradox.' It appears that we might have by contradiction – and without recourse to teleology – the conclusion that the information that drives a PECT describing the Cosmos must have a level of self-correlation, and thus a hyperphysical dimension. We seem to have an account of the origin of agency by means of thermodynamics – and one which seems, *inter alia*, to make variation of species under *domestication* [35] a partial emanation of the *anthrop*ontic. While we *cannot* – and maybe should not try – to characterise *fully* the nature of our personal being, we *can* infer that it is deeply related to the production of *real* structure.

To probe this idea another way, when we try to stop bits rotting in computers by forward error correction, we are preventing extensional *natural*

---

[4] *i.e.*, where the self-information of the hyperkeimenontic and cosmontic are the same.

information from accumulating locally: a kind of informational monocul-
ture, though not necessarily a pernicious one. Order as the exclusion of
natural information may seem a little slippery to pin down, but heat pumps
are not. Is there a heat pump in Nature *not* produced by the agency of humans? We cannot think of one. True: there are endothermic processes, but
heat does not spontaneously move from a cooler body to a hotter, 'unaided
by any external agency.' [146, p. 266] We assume an alien with Cosmic time
on their hands regarding the Earth from afar would observe the advance of
humanity as spontaneous. It is quite delicious that it is a *refrigerator* that Eric
Idle steps out of to expound the wonders of the Cosmos [108]. Did Monty
Python know something the rest of us do not, or was there something they
did *not* know?[5] The more heat pumps we build, the stronger a constructive
proof we seem to have that if we are merely *spontaneously* emergent phe-
nomena from physical noise, something is up. The only way we can obey
the second law extensively is if we skew the data intensionally intentionally.
Applying 'Nature abhors a contradiction,' either the input data of entropy
increases are skewed or the equilibration aspect of the second law is wrong;
we prefer the first explanation. Unless Douglas Adams was largely right
about the dolphins [4, p. 119], we are the only species on Earth to objectify
our own logic and conceptualise this. Such a conclusion seems to place an
awesome responsibility on mankind in the ways it chooses to exercise its
necessarily diabatic agency so as not to do so at the expense of the extant accessible extension of other hyperkeimenontic order in the Cosmos: garbage
in, garbage out. It seems that human reason – powerful as it is – as an account of all there is, destroys itself through the deductive apparatus of its
own logic. On the other hand, the more we conserve the accessibility of the
sublime and add to the *real* structure of the Cosmos – *not* at the expense of
that which already exists – the more we seem to affirm our own humanity.

   This ought not to be very revolutionary: all we have really done is to drill
a little further into the nature of Lord Kelvin's 'external' using our technique
of PECT axiomatisation, which, as we developed in Chapter 2, is little more

---

[5] If we shift from Greek to French through Latin, the hyperkeimenontic can be quite surreal.

than a rephrasing of some ideas of Babbage. We hope this is made more convincing by our sketch of a possible solution to Hilbert's sixth problem in Section 6.1. Our ideas could perhaps be traced back to a divergence from an *apparent* philosophical slip in Rankine's and the Countess of Lovelace's implicit assumption of the nature of a prime mover (though a slip more in the eye of the beholder than the originator): in Rankine's case, to consider that a steam engine *is* one [127] as opposed to being intensionally driven by *information*;[6] in Lady Lovelace's case, to suppose that a computer could compose 'scientific' music [95, p. 694] of any interest once the requirement to input coherent information so as to produce a 'result' is relaxed. Lady Lovelace's stance is far more sophisticated than a superficial interpretation of that observation would suggest. First, there is the development that follows (amplifying Babbage) 'The Analytical Engine has no pretensions whatever to *originate*[7] anything.' [95, p. 722][8] Second, and perhaps more pertinently to the present discussion, she develops a much more interesting position – though largely ignored under the present *Zeitgeist*:

> 'Those who view mathematical science not merely as a vast body of abstract and immutable truths, whose intrinsic beauty, symmetry and logical completeness, when regarded in their connexion together as a whole, entitle them to a prominent place in the interest of all profound and logical minds, but as possessing a yet deeper interest for the human race, when it is remembered that this science constitutes the language through which alone we can adequately express the great facts of the natural world, and those unceasing changes of mutual relationship which, visibly or invisibly, consciously or unconsciously to our immediate physical perceptions, are interminably going on in the agencies of the creation we live amidst...' [95, p. 696]

---

[6] It would be unfair to suggest Rankine himself used that term in a particularly loaded way: he is quite clear on the *amplifying* effect of machines.

[7] Emphasis Lady Lovelace's.

[8] This view has been out of favour since Turing dismissed it [151, pp. 450–451]. On this, we think Turing was *wrong*, and the reasons why ought to be inherent in his own work on decidability and logic [152, 153].

We might perhaps conclude from this that the 'scientific music' Lady Lovelace has in mind is of an altogether more metaphorical kind. Physics seems a loom less of Jacquard or even fair organ – more of life. Steam engines and computers are not *prime* movers; they selectively and transitively *amplify* certain prime move*ments* that remain mysterious.

What is our conclusion? Is all this a candidate theory of everything? No. It is a partial theory of one aspect of reality. A PECT theory is not a PERFECT theory:[9] it says nothing about the structure or fearful symmetry of hyperkeimenontic structure other than proposing constraints within which extensional physical reality appears to affect the information content of future events. We leave the reader to draw their own metaphysical conclusions, but pray pardon a peroration in the paradigm of a Periclean [148, pp. 144–151] panegyric – funnelling some final observations on the physical consequences of free agency as we fuse and flatten the fourth wall in an unapologetically punning paean to paternity. If the great globe itself is a beautiful ship in a squall, it seems in this present to be a ship as imagined by Plato crewed by fools [125, Republic 488c]: now surrounded by light and an ocean of clean fuel they persist – blind to their own shadow – in burning the cabin furniture, wasting the provisions and jettisoning the ship's menagerie, while pursuing titanic quarrels amongst themselves and clogging any route to the lifeboats with an accumulating celestial corona of thorny space flotsam. The explorers who neglect the seaworthiness of their ship seem set to founder before they can discover very much.

If we are grafted into a single lattice, are we to make our patter song an an 'eternal golden braid' [73] of heavenly art and music – or a nightmarish and diabolically cacophonic knot? Can we not tread softly upon such embroidery? Is the ship prematurely to become a muddy vesture of decay? It surely does not need to meet such a fate. Humans can bootstrap machines that can bootstrap machines to perform mechanical tasks: Turing himself observed of programming that '...any processes that are quite mechanical may be turned over to the machine itself.' [150, p. 18] Notwithstanding

---

[9] Nor even practically perfect.

that this insight dates from the dawn of electronic computing, we seem to have lost sight of it in our productivity stasis, while deceiving ourselves in thinking that processes that *cannot* be mechanical[10] can somehow be anthropontically synthesised and treated anthropomorphically as if they are not. Human effort can be applied exponentially and respectfully given the loving application of care and creativity, so why is so much of it spent on the repetitive, the nugatory and the destructive? 'When I consider how my light is spent,' [106] 'O Freunde, nicht diese Töne!' [15, p. 111] It is not even necessary to pump out the heat to refrigerate the ship: pumping out the blanket[11] by our *agency* would suffice. If most of it can be pumped in in less than a hundred years, can it not be pumped out in but a fraction of that? If this ark within an ark, this sphere within a sphere, is not just a ship but our cradle – tossed by mountainous seas – then surely it is within our grown-up talents to tidy up our nursery? Then who knows what sweet larks we will some day hear carolling harmoniously – as we climb to find a *Hirtengesang*, treading dawn lightly from mortal dreams made material?   S. D. G.

---

[10] Have we learnt nothing from endless Pygmalion tropes in fiction?

[11] Inducing a disequilibrium in the distribution of pollutants and reducing the information available to be encoded in the mixing of the atmosphere.

References

[1] Edwin A. Abbott. *Flatland*. Oxford World's Classics, 2006.

[2] J.-R. Abrial. *The B-Book: Assigning programs to meanings*. Cambridge University Press, 1996.

[3] Jean-Raymond Abrial. *Modeling in Event-B*. Cambridge University Press, 2010.

[4] Douglas Adams. *The Hitch Hiker's Guide to the Galaxy*. Pan Books Ltd, 1979.

[5] John D. Anderson et al. "Anomalous Orbital-Energy Changes Observed during Spacecraft Flybys of Earth." In: *Physics Review Letters* 100.9 (Mar. 2008), p. 091102. DOI: 10.1103/PhysRevLett.100.091102.

[6] John D. Anderson et al. "Measurements of Newton's gravitational constant and the length of day." In: *arXiv:1504.06604v2 [gr-qc]* (May 2015). Preprint.

[7] John D. Anderson et al. "Reply to comment by M. Pitkin on 'Measurements of Newton's gravitational constant and the length of day' by Anderson J. D. et al." In: *arXiv:1508.00532v1 [gr-qc]* (July 2015). Preprint.

[8] Aristotle. *Aristotelis Opera, Volume 1*. Ancient Greek. Ed. by Immanuel Bekker and Olof Gigon. De Gruyter, 1960. DOI: 10.1515/9783110835861.

[9] Steve Awodey et al. "Voevodsky's Univalence Axiom in Homotopy Type Theory." In: *Notices of the American Mathematical Society* 60.9 (Oct. 2013), pp. 1164–1167. DOI: 10.1090/noti1043.

[10] Charles Babbage. *On the Economy of Machinery and Manufactures*. Cambridge Library Collection – History of Printing, Publishing and Libraries. Cambridge: Cambridge University Press, 2010. DOI: 10.1017/CBO9780511696374.018.

[11] Charles Babbage. *Passages from the Life of a Philosopher*. Cambridge Library Collection – Technology. Cambridge: Cambridge University Press, 2011. DOI: 10.1017/CBO9781139103671.006.

[12] Julija Bagdonaite et al. "A Stringent Limit on a Drifting Proton-to-Electron Mass Ratio from Alcohol in the Early Universe." In: *Science* 339.6115 (Jan. 2013), pp. 46–48. DOI: 10.1126/science.1224898.

[13] Henk Barendregt and Herman Geuvers. "Proof-Assistants Using Dependent Type Systems." In: *Handbook of Automated Reasoning*. Ed. by Alan Robinson and Andrei Voronkov. Handbook of Automated Reasoning. Amsterdam: North-Holland, 2001. Chap. 18, pp. 1149–1238. DOI: 10.1016/B978-044450813-3/50020-5.

[14] Kent Beck et al. *Manifesto for Agile Software Development*. https://agilemanifesto.org. Accessed: 29 October 2021.

[15] Ludwig van Beethoven. *Sinfonie mit Schluss-Chor über Schillers Ode: „An die Freude"*. Symphony No. 9, Op. 125. Mainz: Schott, 1826.

[16] J. S. Bell. "On the Einstein Podolsky Rosen paradox." In: *Physics Physique Физика* 1.3 (Nov. 1964), pp. 195–200. DOI: 10.1103/PhysicsPhysiqueFizika.1.195.

[17] C. H. Bennett. "Logical Reversibility of Computation." In: *IBM Journal of Research and Development* 17.6 (Nov. 1973), pp. 525–532.

[18] Yves Bertot. "CoInduction in Coq." In: *arXiv:cs/0603119v1* (Mar. 2006).

[19] Antoine Bérut et al. "Experimental verification of Landauer's principle linking information and thermodynamics." In: *Nature* 483 (2012), pp. 187–189. DOI: 10.1038/nature10872.

[20] Dines Bjørner. "The Vienna Development Method: The Meta-Language." In: *Lecture Notes in Computer Science*. Ed. by E. K. Blum, M. Paul, and S. Takasu. Vol. 75. Berlin, Heidelberg, New York: Springer-Verlag, 1979, pp. 326–359.

[21] Luca Bombelli et al. "Space-time as a causal set." In: *Physics Review Letters* 59.5 (Aug. 1987), pp. 521–524. DOI: 10.1103/PhysRevLett.59.521.

[22] Hermann Bondi. *Assumption and Myth in Physical Theory*. Cambridge University Press, 1967.

[23] Hermann Bondi. *Relativity and Common Sense: a New Approach to Einstein*. Dover Publications Inc., 2012.

[24] Max Born and Vladimir Fock. "Beweis des Adiabatensatzes." German. In: *Zeitschrift für Physik* 51 (1928), pp. 165–180. DOI: 10.1007/BF01343193.

[25] Bureau International des Poids et Mesures. *Le Système international d'unités.* French and English. 9e édition 2019.

[26] Bureau International des Poids et Mesures. *Le Système international d'unités.* French and English. 8e édition 2006.

[27] Rudolf Carnap. *Meaning and Necessity: A Study in Semantics and Modal Logic.* Chicago: The University of Chicago Press, 1964.

[28] Sadi Carnot. *Réflexions sur la puissance motrice du feu et sur les machines propres à développer cette puissance.* French. Paris: Bachelier, 1824.

[29] Henry Cavendish. "Experiments to determine the Density of the Earth." In: *Philosophical Transactions of the Royal Society of London* 88 (1798), pp. 469–526. DOI: 10.1098/rstl.1798.0022.

[30] Gregory J. Chaitin. "A Theory of Program Size Formally Identical to Information Theory." In: *Journal of the ACM* 22.3 (July 1975), pp. 329–340. DOI: 10.1145/321892.321894.

[31] George Chapline. "Dark Energy Stars." In: *arXiv:astro-ph/0503200v2* (2005). Revised version of paper given at the Texas Conference on Relativistic Astrophysics, Stanford, December 2004.

[32] Alonzo Church. "An Unsolvable Problem of Elementary Number Theory." In: *American Journal of Mathematics* 58.2 (Apr. 1936), pp. 345–363.

[33] Rudolf Clausius. *The mechanical theory of heat, with its applications to the steam-engine and to the physical properties of bodies.* Ed. by T. Archer Hirst. London: John van Voorst, 1867.

[34] R. T. Cox. "Probability, Frequency and Reasonable Expectation." In: *American Journal of Physics* 14.1 (Jan. 1946–Feb. 1946).

[35] Charles Darwin. *The Origin of Species: By Means of Natural Selection, or the Preservation of Favoured Races in the Struggle for Life.* 6th ed. Cambridge Library Collection – Darwin, Evolution and Genetics. Cambridge: Cambridge University Press, 2009. DOI: 10.1017/CBO9780511694295.

[36] Jared Curran Davis. "A Self-Verifying Theorem Prover." PhD thesis. The University of Texas at Austin, 2009.

[37] Peter Debye. "Zur Theorie der spezifischen Wärmen." German. In: *Annalen der Physik* 344.14 (1912), pp. 789–839. DOI: 10.1002/andp.19123441404.

[38] Daniel C. Dennett. *Consciousness Explained*. Penguin Books, 1993.

[39] René Descartes. "Principles of Philosophy." In: *The Philosophical Writings of Descartes*. Vol. 1. Translated by John Cottingham, Robert Stoothoff and Dugald Murdoch. Cambridge: Cambridge University Press, 1985, pp. 177–292. DOI: 10.1017/CBO9780511805042.007.

[40] D. Deutsch. "Quantum Theory, the Church-Turing Principle and the Universal Quantum Computer." In: *Proceedings of the Royal Society of London. A. Mathematical and Physical Sciences* 400.1818 (1985), pp. 97–117. DOI: 10.1098/rspa.1985.0070.

[41] Paul A. M. Dirac. "The Cosmological Constants." In: *Nature* 139 (Feb. 1937), p. 323.

[42] Paul A. M. Dirac. "The Quantum Theory of the Emission and Absorption of Radiation." In: *Proceedings of the Royal Society of London. Series A* 114.767 (1927), pp. 243–265. DOI: 10.1098/rspa.1927.0039.

[43] Christian Doppler. *Ueber das farbige Licht der Doppelsterne und einiger anderer Gestirne des Himmels*. German. Prague: Borrosch und André, 1842.

[44] Adrien Douady and John H. Hubbard. *Étude Dyanmique des Polynômes Complexes*. French. Documents Mathématiques. Paris: Société Mathématique de France, 2007.

[45] Gilles Dowek et al. *The Coq Proof Assistant User's Guide, version 5.6*. Tech. rep. 134. Institut National de Recherche en Informatique et en Automatique, Dec. 1991.

[46] A. Einstein. "Zur Elektrodynamik bewegter Körper." German. In: *Annalen der Physik* 322.10 (1905), pp. 891–921. DOI: 10.1002/andp.19053221004.

[47] Albert Einstein. "Die Grundlage der allgemeinen Relativitätstheorie." German. In: *Annalen der Physik* 354.7 (1916), pp. 769–822. DOI: 10.1002/andp.19163540702.

[48] Conal Elliott and Paul Hudak. "Functional Reactive Animation." In: *Proceedings of the Second ACM SIGPLAN International Conference on Functional Programming*. ICFP '97. New York, NY: Association for Computing Machinery, 1997, pp. 263–273. DOI: 10.1145/258948.258973.

[49] Euclid. *Euclid's Elements of Geometry*. Ancient Greek and English. Ed. by J. L Heiberg. Translated by Richard Fitzpatrick. Richard Fitzpatrick, 2007.

[50] Robert M. Fano. *Transmission of information: A statistical theory of communications*. Cambridge, Massachusetts: The M.I.T. Press, 1961.

[51] Richard Feynman et al. *The Feynman lectures on physics*. Addison-Wesley, 1963.

[52] Richard P. Feynman. *QED: The Strange Theory of Light and Matter*. Penguin Books, 1990.

[53] Luciano Floridi. *The Philosophy of Information*. Oxford University Press, 2011.

[54] N.E. Fuchs. "Specifications are (preferably) executable." In: *Software Engineering Journal* 7.5 (1992), pp. 323–334.

[55] Robin Gandy. "Church's Thesis and Principles for Mechanisms." In: *The Kleene Symposium*. Ed. by Jon Barwise, H. Jerome Keisler, and Kenneth Kunen. Vol. 101. Studies in Logic and the Foundations of Mathematics. Elsevier North-Holland, Inc., 1980, pp. 123–148. DOI: 10.1016/S0049-237X(08)71257-6.

[56] Samuel R. J. George. "Design, formalization and realization of Harmonic Box Coordination Language: an externally timed specification substrate for arbitrarily reliable distributed systems." PhD thesis. University of Bristol, 2014.

[57] Edmund L. Gettier. "Is Justified True Belief Knowledge?" In: *Analysis* 23.6 (1963), pp. 121–123.

[58] J. Willard Gibbs. *The Scientific papers of J. Willard Gibbs*. London: Longmans, Green, and Co., 1906.

[59] W. S. Gilbert. *H. M. S. Pinafore*. New York: G. W. Carleton & Co., Publishers, 1879.

[60] Kurt Gödel. "On Formally Undecidable Propositions of Principia Mathematica and Related Systems I." In: *The Undecidable*. Ed. by Martin Davis. Corrected republication of 1965 edition of Raven Press Books, Ltd. Translated by E. Mendelson from original German in *Monatshefte für Mathematik und Physik*, vol 38 (1931) pp. 173–198. Dover Publications Inc., 2004, pp. 4–38.

[61] Joseph A. Goguen and Rod M. Burstall. "Institutions: Abstract Model Theory for Specification and Programming." In: *Journal of the ACM* 39.1 (Jan. 1992), pp. 95–146. DOI: 10.1145/147508.147524.

[62] M. Gordon, R. Milner, and C.P. Wadsworth. *Edinburgh LCF: A Mechanized Logic of Computation*. Vol. 78. Lecture Notes in Computer Science. Springer, 1979.

[63] A. Gravell and P. Henderson. "Executing formal specifications need not be harmful." In: *Software Engineering Journal* 11.2 (1996), pp. 104–110.

[64] Matthew Dennis Haines. "Distributed runtime support for task and data management." Also as Technical Report CS-93-1100. Ph.D. Colorado State University, Aug. 1993.

[65] Richard S. Hamilton. "Three-manifolds with positive Ricci curvature." In: *Journal of Differential Geometry* 17.2 (1982), pp. 255–306. DOI: 10.4310/jdg/1214436922.

[66] John Harrison. "Towards self-verification of HOL Light." In: *Proceedings of the third International Joint Conference, IJCAR 2006*. Ed. by Ulrich Furbach and Natarajan Shankar. Vol. 4130. Lecture Notes in Computer Science. Seattle, WA: Springer-Verlag, 2006, pp. 177–191.

[67] Stephen Hawking. "Gravitationally Collapsed Objects of Very Low Mass." In: *Monthly Notices of the Royal Astronomical Society* 152.1 (Apr. 1971), pp. 75–78. DOI: 10.1093/mnras/152.1.75.

[68] Ian Hayes and C. B. Jones. "Specifications are not (necessarily) executable." In: *Softw. Eng. J.* 4.6 (Nov. 1989). Pub. Institution of Electrical Engineers for Institution of Electrical Engineers and British Computer Society, pp. 330–338. DOI: 10.1049/sej.1989.0045.

[69] Werner Heisenberg. "Über den anschaulichen Inhalt der quantentheoretischen Kinematik und Mechanik." German. In: *Zeitschrift für Physik* 43 (Mar. 1927), pp. 172–198.

[70] David Hilbert. "Mathematical problems." In: *Bulletin of the American Mathematical Society* 8.10 (1902), pp. 437–479. DOI: 10.1090/S0002-9904-1902-00923-3.

[71] Albert Hofstadter. "Professor Ryle's Category-Mistake." In: *The Journal of Philosophy* 48.9 (1951), pp. 257–270.

[72] Douglas Hofstadter. *I am a Strange Loop*. Basic Books, 2007.

[73] Douglas R. Hofstadter. *Gödel, Escher, Bach: an Eternal Golden Braid*. Basic Books, 1979.

[74] David Hume. *A Treatise of Human Nature*. Ed. by L. A. Selby-Bigge. Edited, with an analytical index, by L. A. Selby-Bigge. Oxford at the Clarendon Press, 1896.

[75] *IEC Standard 80000-13:2008: Quantities and units – Part 13: Information science and technology*. IEC 80000-13:2008.

[76] International Astronomical Union. *Resolutions of the XXIst General Assembly of the International Astronomical Union*. English and French. Accessed 19 December 2021. 1991. URL: https://www.iau.org/static/resolutions/IAU1991_French.pdf.

[77] *ISO/IEC Standard 13568: Information Technology — Z formal specification notation — Syntax, type system and semantics*. ISO/IEC 13568:2002.

[78] E. T. Jaynes. "The Gibbs Paradox." In: *Maximum Entropy and Bayesian Methods: Seattle, 1991*. Ed. by C. Ray Smith, Gary J. Erickson, and Paul O. Neudorfer. Dordrecht: Springer Netherlands, 1992, pp. 1–21. DOI: 10.1007/978-94-017-2219-3_1.

[79] E. E. Constance Jones. "A New Law of Thought." In: *Proceedings of the Aristotelian Society* 11 (1910), pp. 166–186.

[80] Yonggun Jun, Momčilo Gavrilov, and John Bechhoefer. "High-Precision Test of Landauer's Principle in a Feedback Trap." In: *Physics Review Letters* 113.19 (Nov. 2014), p. 190601. DOI: 10.1103/PhysRevLett.113.190601.

[81] Immanuel Kant. *Critique of Pure Reason*. Ed. by Paul Guyer and Allen W. Wood. The Cambridge Edition of the Works of Immanuel Kant. Cambridge: Cambridge University Press, 1998. DOI: 10.1017/CBO9780511804649.

[82] M. G. Kendall. "On the Reconciliation of Theories of Probability." In: *Biometrika* 36.1/2 (1949), pp. 101–116.

[83] Søren Kierkegaard. *Concluding Unscientific Postscript to the Philosophical Crumbs*. Edited and translated by Alastair Hannay. Cambridge University Press, 2009.

[84] Stephen C. Kleene. *Introduction to Metamathematics*. North Holland, 1980.

[85] Yoshio Koide. "A fermion-boson composite model of quarks and leptons." In: *Physics Letters B* 120.1 (1983), pp. 161–165. DOI: 10.1016/0370-2693(83)90644-5.

[86] A. N. Kolmogorov. *Foundations of the Theory of Probability*. Translation edited by Nathan Morrison. New York: Chelsea Publishing Company, 1950.

[87] Leslie Lamport. "The Temporal Logic of Actions." In: *ACM Trans. Program. Lang. Syst.* 16.3 (May 1994), pp. 872–923. DOI: 10.1145/177492.177726.

[88] Leslie Lamport. "Time, clocks, and the ordering of events in a distributed system." In: *Commun. ACM* 21.7 (1978), pp. 558–565. DOI: 10.1145/359545.359563.

[89] L. D. Landau and E. M. Lifshitz. *The Classical Theory of Fields*. Third Revised English Edition. Course of Theoretical Physics, vol. 2. Translated by Morton Hamermesh. Pergamon Press Ltd, 1971.

[90] Rolf Landauer. "Logical Reversibility of Computation." In: *IBM Journal of Research and Development* 5.3 (July 1961), pp. 183–191. DOI: 10.1147/rd.53.0183.

[91] Pierre-Simon, Marquis de Laplace. *A philosophical essay on probabilities*. Translated by Frederick Wilson Truscott and Frederick Lincoln Emory. John Wiley & Sons, 1902.

[92] Simon Lavington. *Early Computing in Britain: Ferranti Ltd. and Government Funding, 1948–1958*. Springer Nature Switzerland AG, 2019.

[93] Georges Lemaître. "The Beginning of the World from the Point of View of Quantum Theory." In: *Nature* 127 (May 1931), p. 706.

[94] Friedrich Lenz. "The Ratio of Proton and Electron Masses." In: *Physical Review* 82.4 (May 1951), p. 554. DOI: 10.1103/PhysRev.82.554.2.

[95] Augusta Ada King, Countess of Lovelace. "Notes by the Translator to 'Sketch of the Analytical Engine invented by Charles Babbage Esq. from the Bibliothèque Universelle de Genève, No. 82, October 1842'." In: *Scientific memoirs, selected from the transactions of foreign academies of science and learned societies, and from foreign journals*. Ed. by Richard Taylor. Vol. 3. The 'Sketch,' translated by Lady Lovelace, which precedes what she modestly describes as 'notes,' is of L.F. Menabrea. Richard and John E. Taylor, 1843. Chap. 29, pp. 691–731.

[96] Titus Lucretius Carus. *De Rerum Natura, Liber Primus*. Latin. Ed. by J. D. Duff. Cambridge University Press, 1923.

[97] Per Martin-Löf. *Intuitionistic Type Theory: Notes by Giovanni Sambin of a series of lectures given in Padua, June 1980*. Bibliopolis, Naples, 1984.

[98] James Clerk Maxwell. "A Dynamical Theory of the Electromagnetic Field."
      In: *Philosophical Transactions of the Royal Society of London* 155 (1865),
      pp. 459–512. DOI: 10.1098/rstl.1865.0008.

[99] James Clerk Maxwell. *Theory of Heat.* Cambridge Library Collection -
      Physical Sciences. Cambridge: Cambridge University Press, 2011. DOI:
      10.1017/CBO9781139057943.

[100] Dennis D. McCarthy and P. Kenneth Seidelmann. "International Atomic
      Time (TAI)." In: *Time: From Earth Rotation to Atomic Physics.* 2nd ed.
      Cambridge: Cambridge University Press, 2018, pp. 225–248. DOI:
      10.1017/9781108178365.015.

[101] James McConnell. "Whittaker's Correlation of Physics and Philosophy." In:
      Proceedings of the Edinburgh Mathematical Society 11.1 (1958), pp. 57–68. DOI:
      10.1017/S0013091500014383.

[102] John Stuart Mill. *A System of Logic, Ratiocinative and Inductive: Being a
      Connected View of the Principles of Evidence, and the Methods of Scientific
      Investigation.* Vol. 2. Cambridge Library Collection – Philosophy.
      Cambridge: Cambridge University Press, 2011. DOI:
      10.1017/CBO9781139149846.

[103] E. A. Milne. *Relativity, Gravitation and World-Structure.* International Series
      of Monographs on Physics. Oxford, at the Clarendon Press, 1935.

[104] Robin Milner, Joachim Parrow, and David Walker. "A calculus of mobile
      processes, I." In: *Information and Computation* 100.1 (1992), pp. 1–40. DOI:
      10.1016/0890-5401(92)90008-4.

[105] Robin Milner, Joachim Parrow, and David Walker. "A calculus of mobile
      processes, II." In: *Information and Computation* 100.1 (1992), pp. 41–77. DOI:
      10.1016/0890-5401(92)90009-5.

[106] John Milton. "Sonnet 19." In: *The Poetical Works of John Milton.* Ed. by
      Henry John Todd. Vol. 5. London: Bye and Law (printer, for various
      subscribers), 1801, p. 489.

[107] Marvin Minsky. "Steps Toward Artificial Intelligence." In: *Proceedings of the
      IRE* 49.1 (Jan. 1961), pp. 8–30. DOI: 10.1109/JRPROC.1961.287775.

[108] Monty Python. *The Meaning of Life.* [Film] Dir. Terry Jones and Terry
      Gilliam. Elstree: The Monty Python Partnership. 1983.

[109]   Till Mossakowski, Christian Maeder, and Klaus Lüttich. "The
        Heterogeneous Tool Set, Hets." In: *Tools and Algorithms for the Construction
        and Analysis of Systems*. Ed. by Orna Grumberg and Michael Huth. Lecture
        Notes in Computer Science. Berlin, Heidelberg: Springer, 2007,
        pp. 519–522. DOI: 10.1007/978-3-540-71209-1_40.

[110]   Magnus O. Myreen and Jared Davis. "A Verified Runtime for a Verified
        Theorem Prover." In: *Interactive Theorem Proving*. Ed. by Marko Eekelen
        et al. Vol. 6898. Lecture Notes in Computer Science. Springer Berlin
        Heidelberg, 2011, pp. 265–280. DOI: 10.1007/978-3-642-22863-6_20.

[111]   NASA. *Solar System's First Interstellar Visitor Dazzles Scientists*. Ed. by
        Tricia Talbert. https://www.nasa.gov/feature/solar-system-s-first-
        interstellar-visitor-dazzles-scientists. Accessed: 18 December 2021.
        Nov. 2017.

[112]   J. von Neumann. "Probabilistic Logics and the Synthesis of Reliable
        Organisms from Unreliable Components." In: *Automata Studies*. Ed. by
        C. E. Shannon and J. McCarthy. Annals of Mathematical Studies 34.
        Princeton University Press, 1956, pp. 43–98.

[113]   John von Neumann. *First Draft of a Report on the EDVAC*. Tech. rep. Version
        prepared by Michael D. Godfrey. Accessed 20 December 2021. Moore
        School of Electrical Engineering, University of Pennsylvania, June 1945.
        URL: https://web.archive.org/web/20130314123032/http:
        //qss.stanford.edu/~godfrey/vonNeumann/vnedvac.pdf.

[114]   Isaac Newton. *Philosophiae Naturalis Principia Mathematica*. Latin. Third
        edition. W. & J. Innys, 1726.

[115]   Tobias Nipkow, Markus Wenzel, and Lawrence C. Paulson. *Isabelle/HOL: a
        proof assistant for higher-order logic*. Berlin, Heidelberg: Springer-Verlag,
        2002.

[116]   Christine Paulin-Mohring. "Inductive Definitions in the System Coq: Rules
        and Properties." In: *Typed Lambda Calculi and Applications*. Ed. by
        Marc Bezem and Jan Friso Groote. Vol. 664. Lecture Notes in Computer
        Science. Springer Berlin Heidelberg, 1993, pp. 328–345. DOI:
        10.1007/BFb0037116.

[117]   Roger Penrose. "'Golden Oldie': Gravitational Collapse: The Role of
        General Relativity." In: *General Relativity and Gravitation* 34 (2002).
        Reprinted from 1969 paper, pp. 1141–1165. DOI: 10.1023/A:1016578408204.

[118] Roger Penrose. *The Emperor's New Mind*. Revised impression as Oxford Landmark Science, 2016. Oxford University Press, 1989.

[119] Grisha Perelman. "Finite extinction time for the solutions to the Ricci flow on certain three-manifolds." In: *arXiv:math/0307245v1* (July 2003). Preprint.

[120] Grisha Perelman. "Ricci flow with surgery on three-manifolds." In: *arXiv:math/0303109v1* (Mar. 2003). Preprint.

[121] Grisha Perelman. "The entropy formula for the Ricci flow and its geometric applications." In: *arXiv:math/0211159v1* (Nov. 2002). Preprint.

[122] M. Pitkin. "Comment on 'Measurements of Newton's gravitational constant and the length of day' by Anderson J. D. et al." In: *EPL* (*Europhysics Letters*) 111.3 (Aug. 2015), p. 30002. DOI: 10.1209/0295-5075/111/30002.

[123] Max Planck. *The theory of heat radiation*. Translated by Morton Masius. Philadelphia: P. Blakiston's, Son & Co., 1914.

[124] Max Planck. "Über irreversible Strahlungsvorgänge." German. In: *Sitzungsberichte der Königlich Preussischen Akademie der Wissenschaften zu Berlin* (1899), pp. 440–480.

[125] Plato. *Plato in Twelve Volumes*. Ancient Greek. Vol. 6. With English translation by Paul Shorey. Cambridge, MA: Harvard University Press, 1980.

[126] T. Rado. "On Non-Computable Functions." In: *The Bell System Technical Journal* 41.3 (May 1962), pp. 877–884. DOI: 10.1002/j.1538-7305.1962.tb00480.x.

[127] William John Macquorn Rankine. *A Manual of the Steam Engine and other Prime Movers*. Glasgow: Richard Griffin and Company, 1859.

[128] Gonzalo Rodriguez-Pereyra. *Leibniz: Discourse on Metaphysics*. Leibniz from Oxford. Oxford University Press, 2020.

[129] Bertrand Russell. "Letter to Frege." In: *From Frege to Gödel: A Source Book in Mathematical Logic, 1879–1931*. Ed. by Jean van Heijenoort. Harvard University Press, 1967, pp. 124–125.

[130] Bertrand Russell. "On Denoting." In: *Mind* XIV.56 (Oct. 1905), pp. 479–493. DOI: 10.1093/mind/XIV.4.479.

[131]   Gilbert Ryle. "Categories." In: *Proceedings of the Aristotelian Society* 38
        (1937), pp. 189–206.

[132]   Erwin Schrödinger. "An Undulatory Theory of the Mechanics of Atoms
        and Molecules." In: *Phys. Rev.* 28.6 (Dec. 1926), pp. 1049–1070. DOI:
        10.1103/PhysRev.28.1049.

[133]   D. Scott. "Models for Various Type-Free Calculi." In: *Proceedings of the
        Fourth International Congress for Logic, Methodology and Philosophy of Science,
        Bucharest, 1971.* Ed. by Patrick Suppes et al. Vol. 74. Studies in Logic and
        the Foundations of Mathematics. Elsevier, 1973, pp. 157–187. DOI:
        10.1016/S0049-237X(09)70356-8.

[134]   Dana S. Scott and Christopher Strachey. "Towards a Mathematical
        Semantics for Computer Languages." In: *Proceedings of the Symposium on
        Computers and Automata* 21 (1972).

[135]   John R. Searle. "Collective Intentions and Actions." In: *Intentions in
        Communication.* Ed. by Jerry Morgan Philip R. Cohen and
        Martha E. Pollack. MIT Press, 1990.

[136]   C. E. Shannon. "A Mathematical Theory of Communication." In: *Bell
        System Technical Journal* 27.3 (July 1948), pp. 379–423. DOI:
        10.1002/j.1538-7305.1948.tb01338.x.

[137]   C.E. Shannon. "Communication in the Presence of Noise." In: *Proceedings
        of the IRE* 37.1 (1949), pp. 10–21. DOI: 10.1109/JRPROC.1949.232969.

[138]   W. Sieg. "Calculation by Man and Machine: Conceptual Analysis." In:
        *Reflections on the Foundations of Mathematics: Essays in Honor of Solomon
        Feferman.* Ed. by Wilfried Sieg, Richard Sommer, and Carolyn Talcott.
        Lecture Notes in Logic 15. A.K. Peters, Ltd for the Association for Symbolic
        Logic, 2002.

[139]   Crosbie W. Smith. "William Thomson and the Creation of
        Thermodynamics: 1840-1855." In: *Archive for History of Exact Sciences* 16.3
        (1977), pp. 231–288.

[140]   M.H. Sørensen and P. Urzyczyn. *Lectures on the Curry-Howard Isomorphism.*
        Ed. by S Abramsky et al. Vol. 149. Studies in Logic and the Foundations of
        Mathematics. Elsevier, 2006.

[141]   M. E. Szarbo. *The Collected Papers of Gerhard Gentzen.* Amsterdam:
        North-Holland Publishing Company, 1969.

[142] Leó Szilárd. "über die Entropieverminderung in einem thermodynamischen System bei Eingriffen intelligenter Wesen." German. In: *Zeitschrift für Physik* 53 (1929), pp. 840–856. DOI: 10.1007/BF01341281.

[143] Alfred Tarski. "Logic, Semantics, Metamathematics: Papers from 1923-38." In: ed. by John Corcoran. Translated by J. H. Woodger. Edited by John Corcoran. Hackett Publishing, 1983. Chap. VIII: The concept of truth in formalized languages, pp. 152–278.

[144] Telecommunication standardization sector of ITU. *Recommendation ITU-T X.660*. Also published as ISO/IEC 9834-1. July 2011.

[145] Todd A. Thompson et al. "A noninteracting low-mass black hole–giant star binary system." In: *Science* 366.6465 (2019), pp. 637–640. DOI: 10.1126/science.aau4005.

[146] William Thomson. "On the Dynamical Theory of Heat, with numerical results deduced from Mr Joule's Equivalent of a Thermal Unit, and M. Regnault's Observations on Steam." In: *Transactions of the Royal Society of Edinburgh* 20.2 (1853), pp. 261–288. DOI: 10.1017/S0080456800033172.

[147] William Thomson. "The Kinetic Theory of the Dissipation of Energy." In: *Proceedings of the Royal Society of Edinburgh* 8 (1875), pp. 325–334. DOI: 10.1017/S0370164600029680.

[148] Thucydides. *History of the Peloponnesian War*. Translated by Rex Warner. Penguin Books, 1972.

[149] Myron Tribus. *Thermostatics and Thermodynamics: An Introduction to Energy, Information and States of Matter, with Engineering Applications*. Princeton, New Jersey: D. Van Nostrand Company, Inc., 1961.

[150] A. M Turing. Report on 'Proposed Electronic Calculator' (The 'ACE report'). Item AMT/C/32 in the Turing Archive, part of the National Archives DSIR 10/385 and held by the Turing Archive at King's College Cambridge, digitised at https://turingarchive.kings.cam.ac.uk/unpublished-manuscripts-and-drafts-amtc/amt-c-32. 1946.

[151] A. M. Turing. "Computing machinery and Intelligence." In: *Mind* 59.236 (Oct. 1950), pp. 433–460. DOI: 10.1093/mind/LIX.236.433.

[152] A. M. Turing. "On computable numbers, with an application to the Entscheidungsproblem." In: *Proceedings of the London Mathematical Society*. Vol. 42. 2. 1936, pp. 230–265.

[153] A. M. Turing. "Systems of Logic Based on Ordinals." In: *Proceedings of the London Mathematical Society* s2-45.1 (1939), pp. 161–228. DOI: 10.1112/plms/s2-45.1.161.

[154] D. A. Turner et al. "Functional Programs as Executable Specifications." English. In: *Philosophical Transactions of the Royal Society of London. Series A, Mathematical and Physical Sciences* 312.1522 (1984), pp. 363–388.

[155] The Univalent Foundations Program. *Homotopy Type Theory: Univalent Foundations of Mathematics.* Institute for Advanced Study, 2013. URL: https://homotopytypetheory.org/book.

[156] Joan A. Vaccaro and Stephen M. Barnett. "Information erasure without an energy cost." In: *Proceedings of the Royal Society of London. A. Mathematical and Physical Sciences* 467 (2011), pp. 1770–1778. DOI: 10.1098/rspa.2010.0577.

[157] John Archibald Wheeler. *Geons, black holes and quantum foam: a life in physics.* New York, NY: W. W. Norton & Company Inc., 2000.

[158] E. T. Whittaker. "On the Functions which are represented by the Expansions of the Interpolation-Theory." In: *Proceedings of the Royal Society of Edinburgh* 35 (1915), pp. 181–194. DOI: 10.1017/S0370164600017806.

[159] Ernst Zermelo. "Investigations in the foundations of set theory I." In: *From Frege to Gödel: A Source Book in Mathematical Logic, 1879–1931.* Ed. by Jean van Heijenoort. Harvard University Press, 1967, pp. 199–215.

Index

Page numbers in italics refer the reader to diagrams. Theories and ideas are nested under and/or labelled with the name of their originator, except in instances where this approach is counter-intuitive or when the theory is mentioned independently of the originator (e.g. Darwinism). Topics forming part of larger subject areas are grouped under the subject to which they conventionally belong: for example, 'artificial intelligence' appears under 'computation'; 'photons' appears under 'particles and quasi-particles.' Following the main text, where an acronym or initialism precedes its definition, it is alphabetised as it is and broken out in parentheses.

Lightning Source UK Ltd.
Milton Keynes UK
UKHW012143310522
403811UK00002B/38